Collins

2026 GUIDE
to the
NIGHT SKY

Radmila Topalovic and Dominic Ford

Published by Collins
An imprint of HarperCollins Publishers
1 Robroyston Gate,
Glasgow G33 1JN
collins.reference@harpercollins.co.uk
collins.co.uk

HarperCollins Publishers
Macken House
39/40 Mayor Street Upper
Dublin 1
D01 C9W8
Ireland

In association with
Royal Museums Greenwich, the group name for the National Maritime Museum,
Royal Observatory Greenwich, Queen's House and *Cutty Sark*
www.rmg.co.uk

First published 2025

© HarperCollins Publishers 2025
Text © Radmila Topalovic 2025
Diagrams © Dominic Ford 2025
Photographs © see credits page 110

Collins® is a registered trademark of HarperCollins Publishers Ltd

All rights reserved. No part of this publication may be reproduced, stored in a retrieval system, or transmitted, in any form or by any means, electronic, mechanical, photocopying, recording or otherwise without the prior permission in writing of the publisher and copyright owners.

Without limiting the exclusive rights of any author, contributor or the publisher of this publication, any unauthorised use of this publication to train generative artificial intelligence (AI) technologies is expressly prohibited. HarperCollins also exercise their rights under Article 4(3) of the Digital Single Market Directive 2019/790 and expressly reserve this publication from the text and data mining exception.

HarperCollins does not warrant that any website mentioned in this title will be provided uninterrupted, that any website will be error free, that defects will be corrected, or that the website or the server that makes it available are free of viruses or bugs. For full terms and conditions please refer to the site terms provided on the website.

The publishers wish to acknowledge their deep debt of gratitude to two of their late astronomy authors: Storm Dunlop FRAS, FRMetS (1942-2025) for his stewardship of the Guides to the Night Sky and Night Sky Almanac and the enormous contribution of his friend and co-author the late Wil Tirion (1943-2024) for his expert star charts and diagrams. They both leave a valuable and ongoing legacy.

A catalogue record for this book is available from the British Library

ISBN 978 0 00 874768 8

10 9 8 7 6 5 4 3 2 1

Printed in Malaysia by Papercraft

If you would like to comment on any aspect of this book, please contact us at the above address or online.
e-mail: collins.reference@harpercollins.co.uk

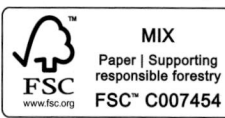

This book contains FSC™ certified paper and other controlled sources to ensure responsible forest management.

For more information visit: www.harpercollins.co.uk/green

Contents

4	Introduction

The Constellations
10	The Southern Circumpolar Constellations
12	The Summer Constellations
13	The Autumn Constellations
14	The Winter Constellations
15	The Spring Constellations

The Moon and the Planets
17	The Moon
18	Map of the Moon
20	Eclipses
22	The Planets
26	Minor Planets
28	Comets
30	Introduction to the Month-by-Month Guide

Month-by-Month Guide
34	January
40	February
46	March
52	April
58	May
64	June
70	July
76	August
82	September
88	October
94	November
100	December

106	Dark Sky Sites
108	Glossary and Tables
110	Acknowledgements
111	Further Information

Introduction

The aim of this Guide is to help people find their way around the night sky at any time of the year, by showing how the stars that are visible change from month to month, and by highlighting various events that occur during 2026. The objects and events described may be observed with the naked eye, or nothing more complicated than a pair of binoculars.

The conditions for observing naturally vary over the course of the year. During the summer, twilight may persist throughout the night and make it difficult to see the faintest stars. There are three recognized stages of twilight: civil twilight, when the Sun is less than 6° below the horizon; nautical twilight, when the Sun is between 6° and 12° below the horizon; and astronomical twilight, when the Sun is between 12° and 18° below the horizon. Full darkness occurs only when the Sun is more than 18° below the horizon. During nautical twilight, only the very brightest stars are visible. During astronomical twilight, the faintest stars visible to the naked eye may be seen directly overhead, but are lost at lower altitudes. As the diagram shows, full darkness persists for about six hours at mid-summer in Sydney. Even at Christchurch, NZ (not shown), full darkness lasts about four hours. By contrast, nautical twilight persists as far south as Cape Horn at mid-summer.

Moonlight will affect the visibility of objects, providing a natural form of light pollution. At Full Moon, it may be very difficult to see some of the fainter stars and objects, and even when the Moon is at a smaller phase it may seriously interfere with visibility if it is near the stars or planets in which you are interested. A full lunar calendar is given for each month and may be used to plan night sky observations. It may also be useful to look at moonrise and moonset times for your particular location.

The celestial sphere

All the objects in the sky (including the Sun, Moon and stars) appear to lie at some indeterminate distance on a large sphere, centred on the Earth. This *celestial sphere* has various reference points and features that are related to those of the Earth. If the Earth's rotational axis (an imaginary line through the North and South Poles and the core of the Earth) is extended, for example, it points to the North and South Celestial Poles. Similarly, the *celestial equator* lies in the same plane as the

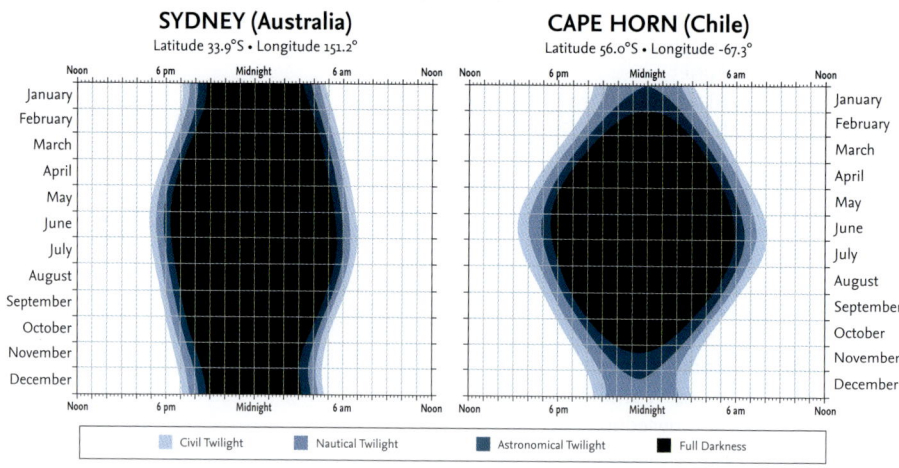

The duration of twilight throughout the year at Sydney and Cape Horn.

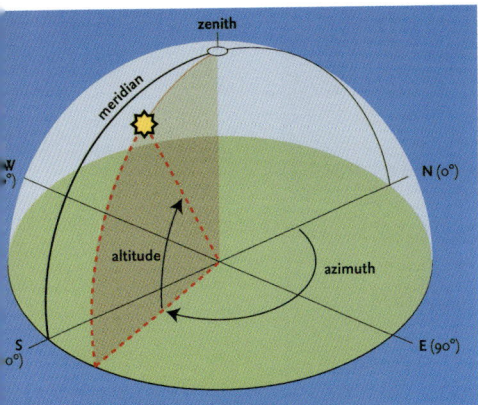
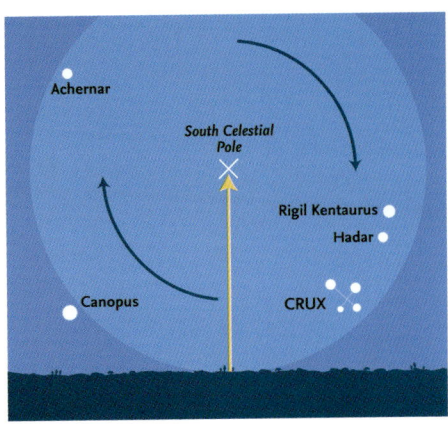

Measuring altitude and azimuth on the celestial sphere.

The altitude of the South Celestial Pole equals the observer's latitude.

Earth's equator; it divides the sky into northern and southern hemispheres. This Guide is written for use in the southern hemisphere, therefore the area of the sky that it describes includes the whole of the southern celestial hemisphere and those portions of the northern sky that become visible at different times of the year. Stars in the far north, however, remain below the horizon throughout the year, and are not included.

It is useful to know some of the astronomy terms for various parts of the sky. As seen by an observer, half of the celestial sphere is invisible at any point in time; these objects will be below the horizon. The point directly overhead is known as the **zenith**, and the (invisible) one below one's feet as the **nadir**. The line running from the north point on the horizon, up through the zenith and then down to the south point is the **meridian**. This is an important invisible line in the sky, because objects are highest in the sky, and thus easiest to see, when they cross the meridian in the south. Objects are said to **transit** when they cross this line in the sky.

In this book, reference is frequently made in the text and in the diagrams to the standard compass points around the horizon. The position of any object in the sky specific to the observer's location on Earth may be described by its **altitude** (measured in degrees above the horizon), and its **azimuth** (measured in degrees from north 0°, through east 90°, south 180° and west 270°). Experienced amateurs and professional astronomers also use another system of specifying locations on the celestial sphere called right ascension and declination, but here the simpler method will suffice.

The celestial sphere appears to rotate about an invisible axis, running between the North and South Celestial Poles. The location (i.e. the altitude) of the Celestial Poles depends entirely on the observer's latitude on Earth. At the South Pole (latitude -90° or 90°S), the South Celestial Pole (SCP) would be directly overhead (at the zenith, or at an altitude of 90°). The charts in this book are produced for the latitude of 35°S, so the SCP is 35° above the southern horizon. The fact that the SCP is fixed relative to the horizon means that all the stars within 35° of the pole are always above the horizon and may, therefore, always be seen at night, regardless of the time of year. This southern circumpolar region is an ideal place to begin learning the sky, and ways to identify the circumpolar stars and constellations will be described shortly.

The ecliptic and the zodiac

Another important line on the celestial sphere is the Sun's apparent path against the background stars – in reality the result of the

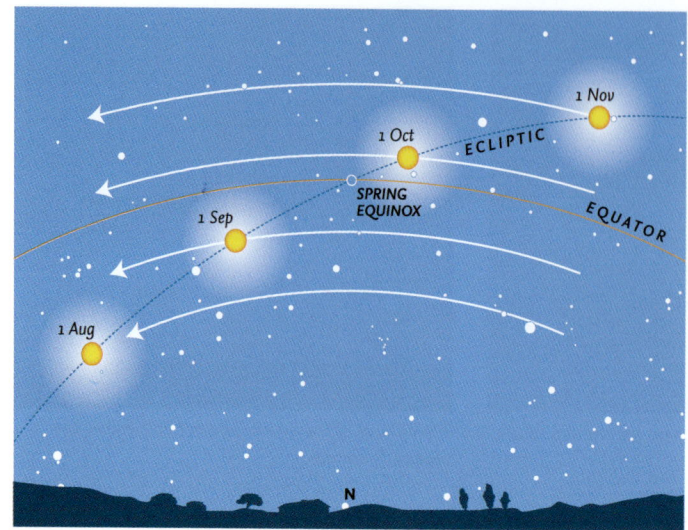

The Sun crossing the celestial equator at the September equinox (spring equinox in the southern hemisphere).

Earth's orbit around the Sun. This is known as the *ecliptic*. The point where the Sun, apparently moving along the ecliptic, crosses the celestial equator from south to north is known as the (southern) autumnal equinox, which occurs on 20 or 21 March. At this time and at the (southern) spring equinox, on 22 or 23 September (when the Sun crosses the celestial equator from north to south), day and night are almost exactly equal in length. The autumnal equinox is currently located in the constellation of Pisces; this means the Sun is moving through Pisces on this day. It is important in astronomy because it defines the zero point for the system of celestial coordinates (right ascension) not used in this Guide.

The Moon and planets are to be found in a band of sky all the way around the celestial sphere that extends 8° on either side of the ecliptic. This is because the orbits of the Moon and planets are inclined at various angles to the ecliptic (i.e. to the plane of the Earth's orbit). This band of sky is known as the zodiac and, when originally devised, consisted of twelve **constellations**, all of which were considered to be exactly 30° wide. When the constellation boundaries were formally established by the International Astronomical Union in 1930, adjustments were made resulting in the ecliptic passing through thirteen constellations and the Moon and planets passing through several other constellations that are adjacent to the original twelve.

The constellations

Since ancient times, the celestial sphere has been divided into various constellations, most dating back to antiquity and usually associated with certain myths or legendary people and animals. The boundaries of the 88 constellations across the whole celestial sphere have been fixed by international agreement and their names (in Latin) are largely derived from Greek or Roman originals. Some of the names of the most prominent stars are of Greek or Roman origin, but many are derived from Arabic names. Many bright stars have no individual names and, for many years, stars were identified by terms such as 'the star in Hercules' right foot'. A more sensible scheme was introduced by the German astronomer Johannes Bayer in the early seventeenth century. Following his scheme – which is still used today – most of the brightest stars are identified by a Greek letter followed by the genitive form of the constellation's Latin name. An example is the Pole Star, also known as Polaris and α Ursae Minoris (abbreviated α UMi). The Greek alphabet is shown on page

110, with a list of all the constellations that may be seen from latitude 35°S on page 109, together with abbreviations, their genitive forms and English names. Other naming schemes exist for fainter stars; these are not used in this book.

Asterisms

Apart from the constellations, certain groups of stars, which may form a part of a larger constellation or cross several constellations, are readily recognizable and have been given individual names. These groups are known as *asterisms*, and the most famous (and well-known) seen from the southern hemisphere is 'the Southern Cross', the common name for the four brightest stars in the constellation of **Crux**. The names and details of some asterisms mentioned in this book are given in the list on page 110.

Magnitudes

The brightness of a star, planet or other body is given in magnitudes (mag.). This is a mathematically defined scale where larger (positive) numbers indicate a fainter object. The scale extends beyond the zero point to negative numbers for very bright objects. (Sirius, the brightest star in the night sky is mag. -1.4.) Most observers are able to see stars as faint as about mag. 6, under very clear skies.

The Moon

Although the daily spin of the Earth carries the sky from east to west (objects in the south rise in the east and set in the west) the Moon gradually moves eastwards relative to the background stars by approximately its diameter in an hour. This is equivalent to about half a degree across the sky. Its apparent eastward motion and change in phase are caused by its orbit around the Earth and changing position relative to the Sun.

Normally, in its orbit around the Earth, the Moon passes above or below the direct line between the Earth and the Sun (at New Moon) or outside the area obscured by the Earth's shadow (at Full Moon), explained in the diagram on page 17. Occasionally, however, the three bodies are more-or-less perfectly aligned to give an *eclipse*: a solar eclipse at New Moon or a lunar eclipse at Full Moon. Depending on the exact circumstances, a solar eclipse may be merely partial (when the Moon does not cover the whole of the Sun's disk); annular (when the Moon is too far from Earth in its orbit to appear large enough to hide the whole of the Sun); or total. Total and annular eclipses are visible from very restricted areas of the Earth, but partial eclipses are normally visible over a wider area.

Somewhat similarly, at a lunar eclipse, the Moon may pass through the lighter outer zone of the Earth's shadow, the **penumbra**, called a penumbral eclipse, which is not generally perceptible to the naked eye; part of the Moon could pass within the darkest part of the Earth's shadow, the **umbra**, in a partial eclipse; or it could move completely within the umbra in a total eclipse. Unlike solar eclipses, lunar eclipses are visible from large areas of the Earth.

Occasionally, as it moves across the sky, the Moon passes between the Earth and individual planets or distant stars, giving rise to an *occultation*. As with solar eclipses, such occultations are visible from restricted areas of the world.

The planets

The planets are always moving against the background stars, therefore they are treated in some detail in the monthly pages and information is given regarding when they are close to the Sun, other planets, the Moon or any of five bright stars that lie near the ecliptic. Such events are known as *appulses* or, more frequently, as *conjunctions*. (There are technical differences in the way these terms are defined – and should be used – in astronomy, but these need not concern us here.) The positions of the planets are shown for every month on a special chart of the ecliptic.

The conditions of most favourable visibility depend on whether the planet is one of the two known as *inferior planets* – planets that orbit the Sun within the Earth's orbit (Mercury and Venus) – or one of the five *superior planets* – planets that orbit the Sun beyond the Earth.

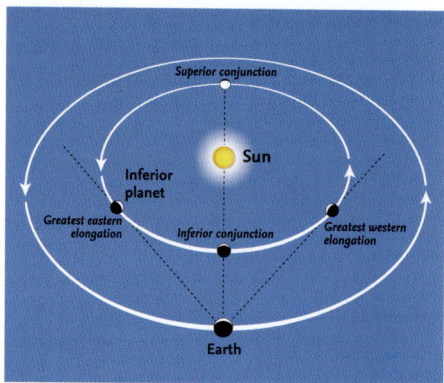

Inferior planet.

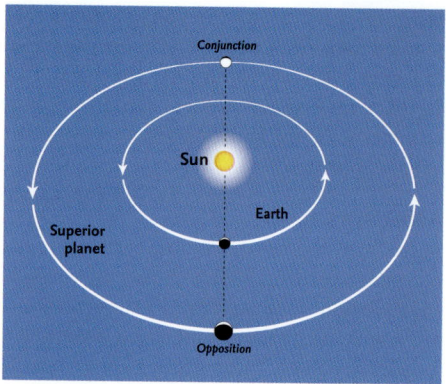

Superior planet.

Of the latter, three (Mars, Jupiter and Saturn) are covered in detail; these are visible to the naked eye. Occasionally, details of the fainter superior planets, Uranus and Neptune, are included, and special charts are given for them on page 25.

The inferior planets are most readily seen at eastern or western elongation, when their angular distance from the Sun (separation from the Sun in degrees) is greatest. Superior planets are best seen at **opposition**, when they are directly opposite the Sun in the sky and cross the meridian at local midnight.

Angular distance can be measured approximately by holding one hand at arm's length. The various angles are shown in the diagram, together with the separations of the various stars in and around Orion.

Meteors

At some time or other, nearly everyone has seen a meteor – a 'shooting star' – as it flashed across the sky. The particles that cause meteors – known technically as 'meteoroids' – normally range in size from that of a grain of sand (or even smaller) to the size of a pea. **Fireballs** or **bolides** are very bright meteors (brighter than mag. -4) that are caused by objects up to 1 metre in size. Fireballs sometimes cause sonic booms that may be heard some time after the meteor is seen. On any night of the year there are occasional meteors, known as sporadics, that may travel in any direction. These occur at a rate that is normally between three and eight per hour.

Far more important, however, are meteor showers, which occur at fixed periods of the year, when the Earth encounters a trail of particles left behind by a comet or, very occasionally, by a minor planet (asteroid). Meteors always appear to diverge from a single point on the sky, known as the radiant, and the radiants of major showers are shown on the charts. Meteors that come from a circular area 8° in diameter around the radiant are classed as belonging to the particular shower. All others that do not come from that area are sporadics (or, occasionally from another shower that is active at the same time). A list of the major meteor showers is given on page 31. Examples of meteors are shown on pages 32, 77 and 101.

Looking directly at the radiant is not the most effective way of seeing meteors. They are most likely to be noticed if one is looking about 40–45° away from the radiant position. This is approximately two hand-spans as shown in the diagram for measuring angles.

Other objects

In the late eighteenth century Charles Messier, a French astronomer, compiled a catalogue of objects while he was searching for comets. These nebulae, clusters of stars and galaxies, were given 'Messier numbers' (some already had names given several thousand years ago such as Praesepe – the Beehive Cluster). Some, such as the Andromeda Galaxy, M31, and the Orion Nebula, M42, may be seen by

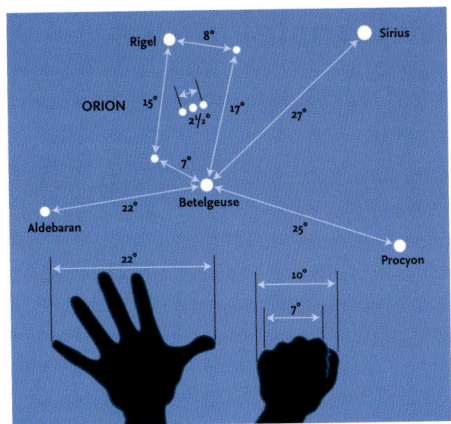

Measuring angles in the sky.

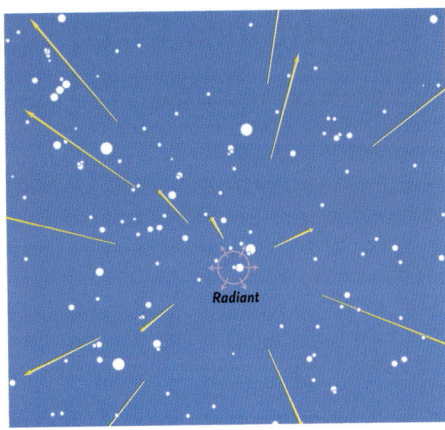

Meteor shower (showing the Geminids radiant).

the naked eye, but all those given in the list will benefit from the use of binoculars. Apart from galaxies, such as M31, which contain thousands of millions of stars, there are also two types of cluster: open clusters, such as M45, the Pleiades, which may consist of a few dozen to some hundreds of stars; and globular clusters, such as Omega Centauri, which are spherical concentrations of many thousands of stars. One or two gaseous nebulae, consisting of gas illuminated by stars within them, are also visible. The Orion Nebula, M42, is one, and is illuminated by a group of four stars, known as the Trapezium, which may be seen by using a good pair of binoculars.

Some interesting objects.

Messier / NGC	Name	Type	Constellation	Maps (months)
—	47 Tucanae	globular cluster	Tucana	All year
—	Hyades	open cluster	Taurus	Nov–Feb
—	Melotte 111 (Coma Cluster)	open cluster	Coma Berenices	Mar–Jun
M3	—	globular cluster	Canes Venatici	Apr–Jul
M4	—	globular cluster	Scorpius	Mar–Sep
M8	Lagoon Nebula	gaseous nebula	Sagittarius	Apr–Oct
M11	Wild Duck Cluster	open cluster	Scutum	May–Oct
M13	Hercules Cluster	globular cluster	Hercules	Jun–Jul
M15	—	globular cluster	Pegasus	Jul–Nov
M20	Trifid Nebula	gaseous nebula	Sagittarius	Apr–Oct
M22	—	globular cluster	Sagittarius	May–Oct
M27	Dumbbell Nebula	planetary nebula	Vulpecula	Jun–Oct
M31	Andromeda Galaxy	galaxy	Andromeda	Oct–Nov
M35	—	open cluster	Gemini	Dec–Mar
M42	Orion Nebula	gaseous nebula	Orion	Oct–Apr
M44	Praesepe	open cluster	Cancer	Jan–Apr
M45	Pleiades	open cluster	Taurus	Nov–Feb
M57	Ring Nebula	planetary nebula	Lyra	Jun–Sep
M67	—	open cluster	Cancer	Jan–May
IC 2602	Southern Pleiades	open cluster	Carina	All year
NGC 2070	Tarantula Nebula	emission nebula	Dorado (LMC)	All year
NGC 3242	Ghost of Jupiter	planetary nebula	Hydra	Jan–Jun
NGC 3372	Eta Carinae Nebula	gaseous nebula	Carina	All year
NGC 4755	Jewel Box	open cluster	Crux	Jan–Sep
NGC 5139	Omega Centauri	globular cluster	Centaurus	Jan–Aug

The Southern Circumpolar Constellations

The southern circumpolar stars are the key to starting to identify the constellations. For anyone in the southern hemisphere they are visible at any time of the year, and nearly everyone is familiar with the striking pattern of four stars that make up the constellation of **Crux** (the Southern Cross) and also the two nearby stars **Rigil Kentaurus** and **Hadar** (α and β Centauri, respectively). Both stars are actually triple star systems, one of the three stars of Rigil Kentaurus is called Proxima Centaurus; it is the closest star to Earth, lying 4.2 light-years away. This star has an Earth-sized planet, which orbits the star every 11.2 days. These circumpolar stars are visible throughout the year for most observers, although for observers farther north, the stars may become difficult to see as they are low on the horizon in the months of October and November.

Crux
The distinctive shape of the constellation of **Crux** is usually easy to identify, although some people (especially northerners unused to the southern sky) may wrongly identify the slightly larger False Cross, formed by the stars **Aspidiske** and **Avior** (ι and ε Carinae respectively) plus **Markeb** and **Alsephina** (κ and δ Velorum). The **Coalsack Nebula** (a vast dark cloud of obscuring dust up to 35 light-years across) is readily visible on the southern side of Crux between **Acrux** and **Mimosa** (α and β Crucis).

A line through **Gacrux** (γ Crucis) and **Acrux** (α Crucis) points approximately in the direction of the South Celestial Pole, which lies in the faint constellation of **Octans**, crossing the faint constellations of **Musca** and **Chamaeleon**. The basic triangular shape of Octans itself is best found by extending a line from **Peacock** (α Pavonis) in the constellation **Pavo**, through β Pavonis, by about the same distance as that between the stars.

Centaurus
Although Crux is a distinctive shape, **Centaurus** is a large, rather straggling constellation, with one notable object: the giant, bright globular cluster **Omega Centauri** (the brightest and largest globular in the sky consisting of around 10 million stars), which lies towards the north, on the line from Hadar (β Centauri) through ε Centauri. A more accurate method of locating the South Celestial Pole is to use the line from Crux and imagine a line at right angles and midway to the line between Hadar and Rigil Kentaurus – the intersection of these two lines identifies the SCP (see the diagram opposite).

Carina
Apart from the two stars that form part of the False Cross, the constellation of **Carina** is, like Centaurus, a large, sprawling constellation. It contains one striking open cluster, the **Southern Pleiades**, and a remarkable emission nebula, the **Eta Carinae Nebula**. The second brightest star in the sky (after Sirius) is **Canopus**, α Carinae, which lies far away to the west.

The Magellanic Clouds
On the opposite side of the South Celestial Pole to Crux and Centaurus lie the two Magellanic Clouds, small irregular satellite galaxies of our Milky Way. The **Small Magellanic Cloud** (SMC) lies a distance of 200,000 light-years from Earth; it sits to one side of the triangular constellation of **Hydrus**,

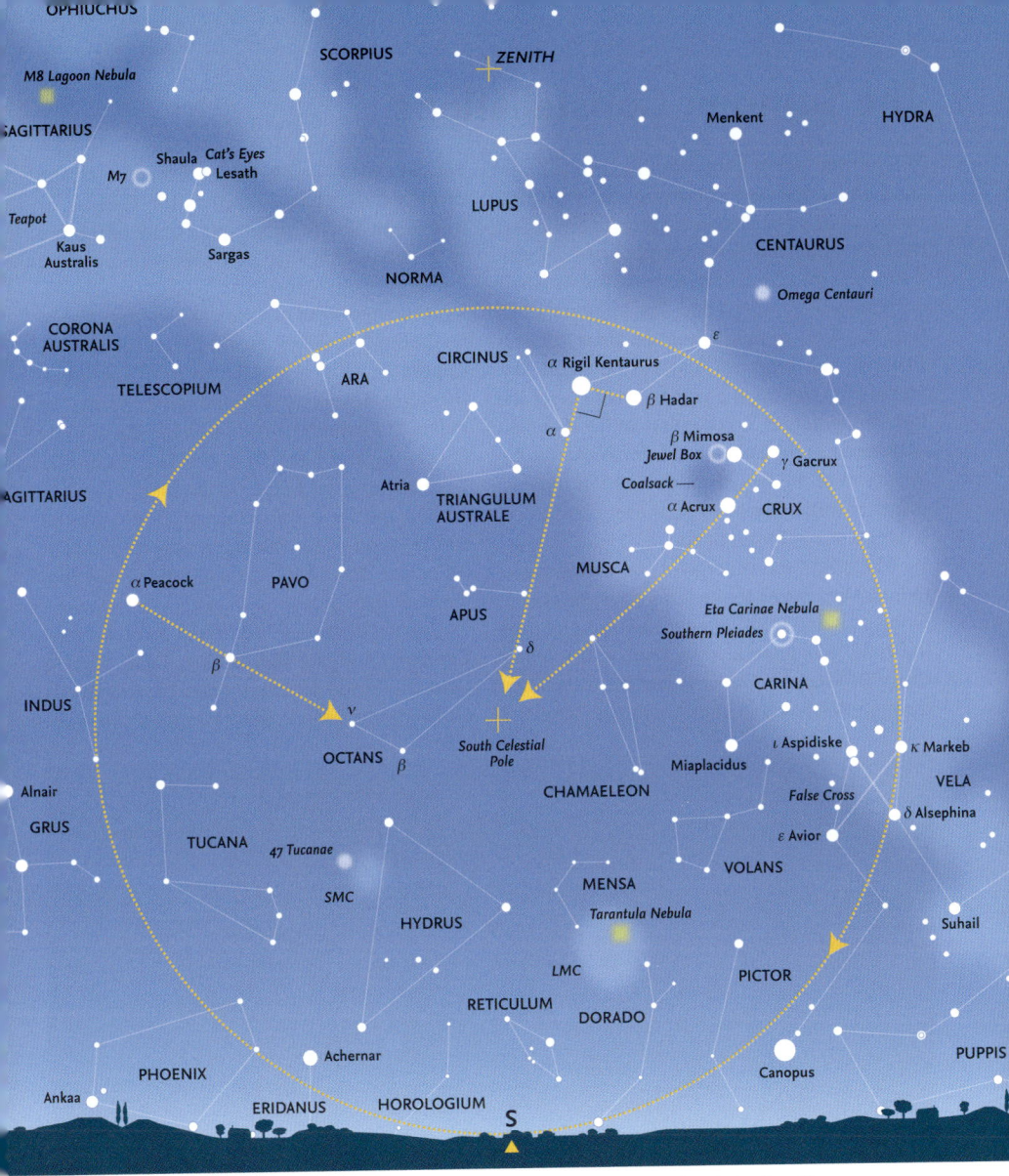

The stars and constellations inside the circle are always above the horizon, seen from latitude 35°S.

but it is actually within the constellation of **Tucana**.

Nearby is another bright globular cluster, **47 Tucanae**. Hydrus itself is also easily identified from the star **Achernar**, α Eridani, the rather isolated brilliant star at the southern end of **Eridanus**, which wanders a long way south, having begun at the foot of Orion.

The Large Magellanic Cloud (LMC) is around 40,000 light-years closer to Earth than the SMC and it lies within the faint constellation of **Dorado**. The LMC contains the large, readily visible **Tarantula Nebula**, an emission nebula that is a major star-forming region. The LMC has an apparent size of around 10° in the sky, equivalent to 20 Moons across.

THE CONSTELLATIONS 11

The Summer Constellations

The summer sky is dominated by several bright stars and distinctive constellations. The most conspicuous constellation is **Orion** (the Hunter), the main body of which has an hourglass shape. It straddles the celestial equator and is thus visible from anywhere in the world. The three stars that form the 'Belt' of Orion point to **Sirius** (α Canis Majoris), the brightest star in the night sky. **Mintaka** (δ Orionis), the westernmost star of the Belt, farthest from Sirius, points in the direction of **Achernar** (α Eridani). A line from **Alnitak** (at the other end of the Belt) through **Saiph** (κ Orionis) points towards **Canopus** (α Carinae), the second brightest star in the night sky.

The bright red supergiant star **Betelgeuse** (α Orionis) lies below the belt, as does **Bellatrix** (γ Orionis). A line from Betelgeuse through Sirius indicates the general direction of the 'False Cross' in **Vela** and **Carina**. A line from Bellatrix, through **Aldebaran** (α Tauri), past the 'V' of the **Hyades** cluster, points to the distinctive cluster of bright blue stars known as the **Pleiades**, or 'the Seven Sisters'. Aldebaran is one of the five bright stars that may sometimes be occulted (hidden) by the Moon.

From the head of Orion, a line from **Meissa** (λ Orionis) through **Elnath** (β Tauri) leads to **Capella** (α Aurigae), and one from **Betelgeuse** through **Alhena** (γ Geminorum) points to **Pollux** (β Geminorum) of **Gemini**. The slightly fainter second head of the twins, Castor (α Geminorum), lies to the northwest.

On the far eastern side of Orion sits the constellation of **Leo** (the lion), easy to find with its 'upside-down backwards question mark' (the 'head' of the lion).

Six bright stars in six different constellations: **Capella** (α Aurigae), **Aldebaran** (α Tauri), **Rigel** (β Orionis), Sirius (α Canis Majoris), **Procyon** (α Canis Minoris) and Pollux (β Geminorum) form

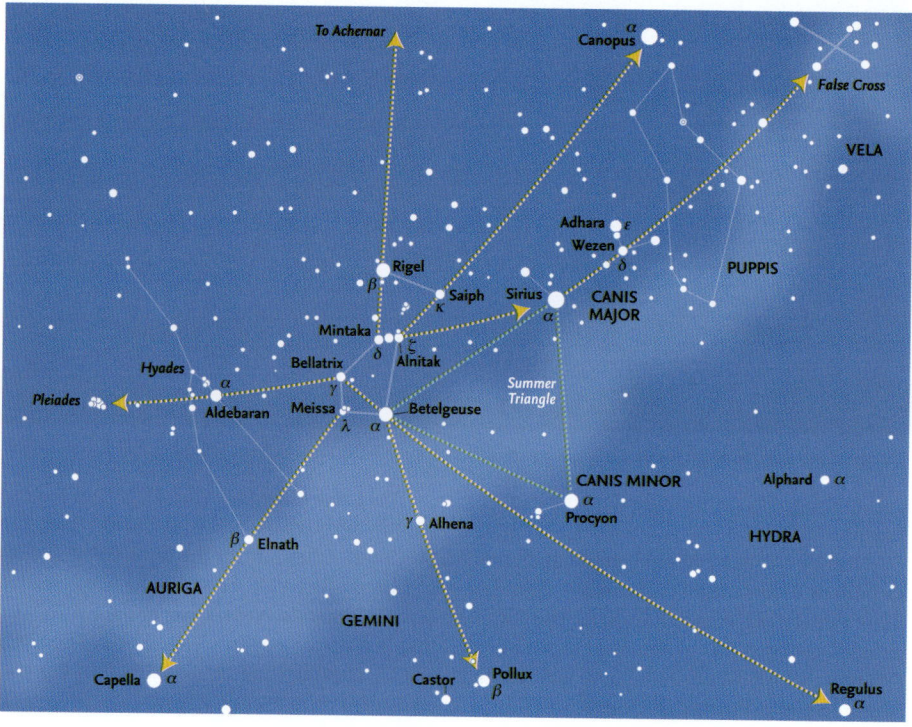

what, for northern observers, is sometimes known as 'the Winter Hexagon'. An almost perfect equilateral triangle, 'the (southern) Summer Triangle', is formed by Betelgeuse (α Orionis), Sirius (α Canis Majoris) and Procyon (α Canis Minoris).

Several of the stars in this region of the sky show distinctive tints: Betelgeuse (α Orionis) is reddish, Aldebaran (α Tauri) is orange and Rigel (β Orionis) is blue-white.

The Autumn Constellations

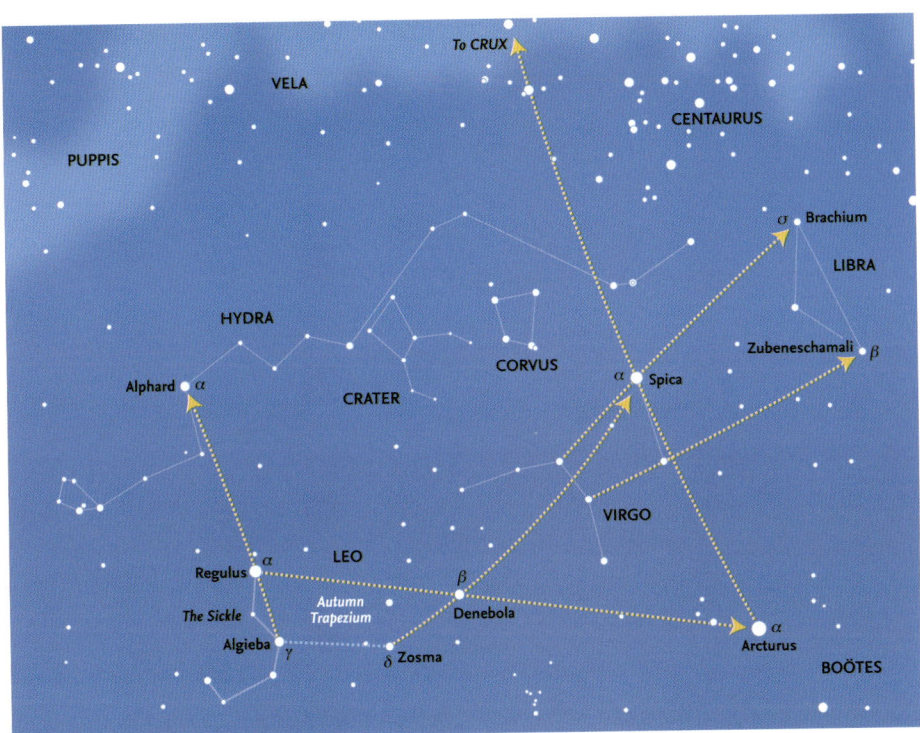

The most prominent constellation in the autumn sky is the zodiacal constellation of **Leo**, and its brightest star, **Regulus** (α Leonis). Regulus is one of the five bright stars that could be occulted by the Moon. Regulus forms the 'dot' of the 'upside-down backwards question mark' of stars, known as 'the Sickle'.

A line extended from **Algieba** (γ Leonis), the second star in 'the Sickle', through Regulus points to **Alphard** (α Hydrae) the brightest star in **Hydra**, the largest of the 88 constellations, which sprawls a long way across the sky from its distinctively shaped 'head' of stars. A line from Regulus through **Denebola** (β Leonis) at the other end of the constellation of Leo points towards **Arcturus** (α Boötis), the brightest star in the northern hemisphere of the sky. The constellation of **Boötes** is sometimes described as 'kite-shaped' or 'shaped like the letter P'. Another line, down the 'back' of Leo, from **Zosma** (δ Leonis) through Denebola points in the general direction of **Spica** (α Virginis) the principal star in the constellation of **Virgo**, a rough quadrilateral of moderately bright stars and fainter lines of stars extending outwards.

The two sides of the main quadrilateral,

THE CONSTELLATIONS 13

if extended, point eastwards towards the two principal stars, **Zubeneschamali** and **Brachium** (β and σ Librae, respectively), of the small zodiacal constellation of **Libra**. The two small constellations of **Corvus** and **Crater** lie between Virgo and the long stretch of Hydra. A line from Arcturus through Spica, if extended far to the south across the sky, points to the distinctive constellation of **Crux**.

The Winter Constellations

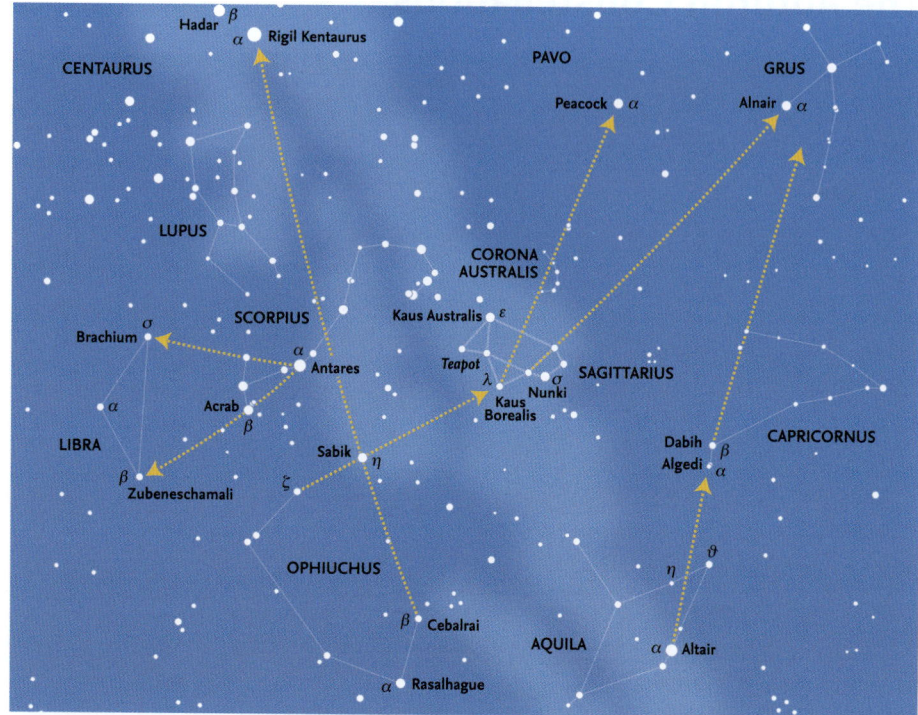

During the winter season, the Milky Way runs right across the sky, and the distinctive constellation of **Scorpius** is clearly visible, with bright red **Antares** (α Scorpii) and the constellation's 'sting' trailing behind it. The two sides of the 'fan' of stars immediately east of Antares, if extended, point to the two principal stars of Libra, **Zubeneschamali** and **Brachium** (β and σ Librae, respectively) on the southern and northern sides of the constellation.

North of Scorpius lies the large constellation of **Ophiuchus**. The ecliptic runs through the southern portion of Ophiuchus, and the Sun spends far more time in that constellation than it does in neighbouring Scorpius. A line from **Cebalrai** (β Ophiuchi) in the north of the constellation, through **Sabik** (η Ophiuchi), if extended right across Scorpius, carries you to the two bright stars of **Centaurus**, **Rigil Kentaurus** and **Hadar** (α and β Centauri, respectively). Between Libra and Centaurus lies the constellation of **Lupus** (the Wolf).

A line from the fainter star ζ Ophiuchi through **Sabik** points towards **Kaus Borealis** (λ Sagittarii) the 'lid' of 'the Teapot' of **Sagittarius**. From there two lines radiate (on the west) across **Corona Australis** towards **Peacock** (α Pavonis) and (on the east) **Alnair** (α Gruis) in the small constellation of **Grus**.

Also north of **Scorpius** is the constellation of **Aquila**, with its brightest star, **Altair** (α Aquilae). The diamond shape of the 'wings' of Aquila may be used to locate **Algedi** (α Capricorni), a prominent double star in **Capricornus**. A line extended through **Dabih** (β Capricorni) and the western side of Capricornus, also points in the direction of Grus.

Aquila (the Eagle) and the constellation of **Cygnus** (the Swan) represent birds 'flying' down the length of the Milky Way. This part of the Milky Way contains the Great Dark Rift, an elongated dark region where the light from distant stars is obscured by intervening dust. The dark Rift is clearly visible even to the naked eye in a dark sky region.

Aquila consists of a diamond shape of stars, representing the body and wings of the eagle, together with a rather faint star, λ Aquilae, marking the 'head'.

The Spring Constellations

In spring, the constellation of **Pegasus** and 'the Great Square of Pegasus' are prominent in the north. However, the star at the northeastern corner, **Alpheratz**, is actually α Andromedae, belonging to the adjacent constellation of **Andromeda**. Between Andromeda and **Cetus** lie the two small constellations of **Triangulum** (the Triangle) and **Aries** (the Ram). A diagonal line across the Square from Alpheratz (α Andromedae) through **Markab** (α Pegasi) points to **Sadalmelik** (α Aquarii), just to the west of the 'Y-shaped' asterism of 'the Water Jar' in the zodiacal constellation of **Aquarius**. Further extended, that line carries the observer to the centre of the triangular constellation of **Capricornus**.

The constellation of Pisces consists of two lines of stars converging to the east of Pegasus. In mythological representations, the two fish are linked by ribbons, tied together at Alrescha (α Piscium). The distinctive asterism of 'the Circlet' lies at the western end of the southern line of stars. The western line of the Great Square passing from **Scheat** (β Pegasi) through Markab, crosses Aquarius and points to the bright, fairly isolated star, Fomalhaut (α Piscis Austrini) in the constellation of Piscis Austrinus (the Southern Fish) and then onward to Tiaki (β Gruis) in the centre of a cross of stars within **Grus**. The eastern line, from Alpheratz through Algenib (γ Pegasi), extended outwards indicates Diphda (β Ceti), which is actually the brightest star in Cetus and further across, Ankaa (α Phoenicis) in the constellation of Phoenix.

The line between Markab and Algenib may be extended as an arc to lead to Menkar (α Ceti) in the 'tail' of Cetus. Diphda and Fomalhaut form the base of an almost perfect, large isosceles triangle, with Achernar (α Eridani) at the other apex. Eridanus itself is a long, trailing constellation that begins far to the north, near Rigel in Orion. West of Rigel lies the prominent cluster of the Pleiades, in the constellation of Taurus.

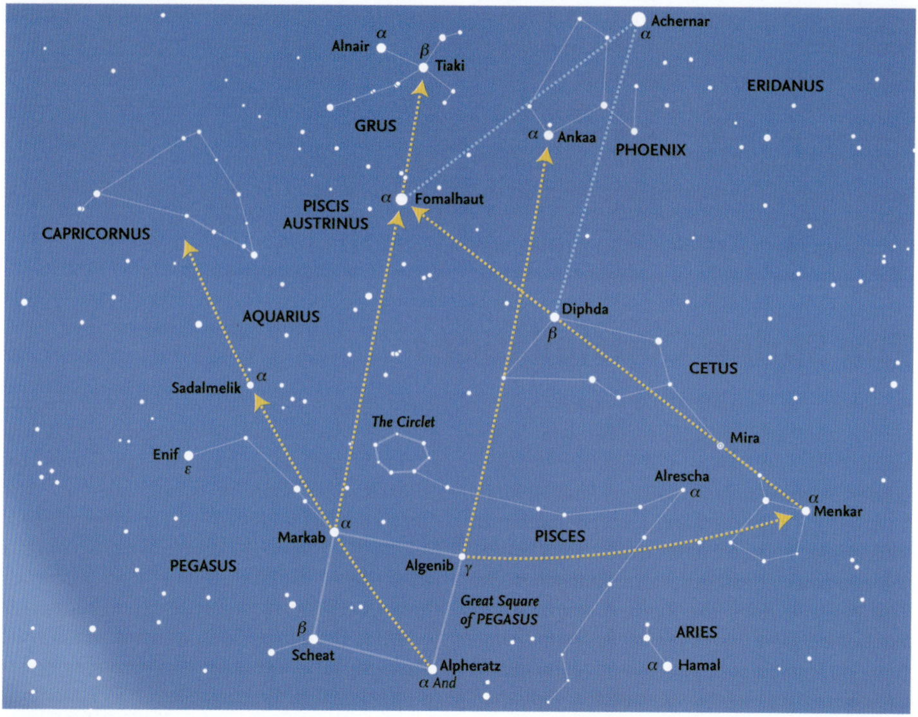

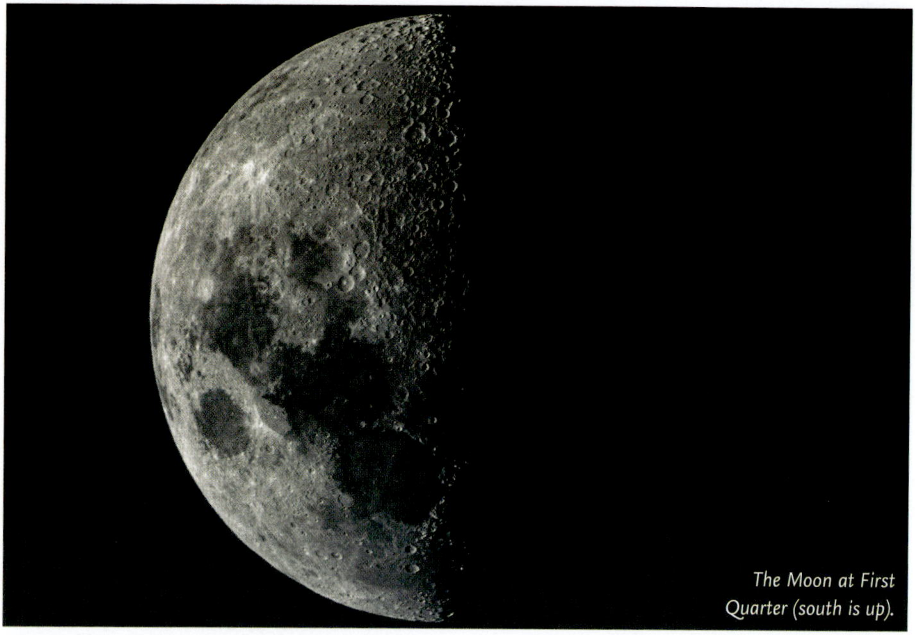

The Moon at First Quarter (south is up).

The Moon

The lunar phase cycle has a duration of 29.5 days and starts with the New Moon, when the near side of the Moon facing us is not illuminated by the Sun. The New Moon rises several hours after midnight and sets in the late afternoon or night; the First Quarter Moon rises in the late morning and sets around midnight; the Full Moon is above the horizon most of the night and the Last Quarter Moon rises around midnight and sets in the late morning.

Although the main features of the surface – the light highlands and the dark maria (seas) – may be seen with the naked eye, far more features may be detected with the use of binoculars or any telescope. The many craters are best seen when they are close to the **terminator** (the boundary between the illuminated and the non-illuminated areas of the surface), when the Sun rises or sets over any particular region of the Moon and the crater walls or central peaks cast strong shadows. These are best seen at crescent, quarter and gibbous phases. Most features become difficult to see at Full Moon, although this is the best time to see the bright ray systems surrounding certain craters. Accompanying the Moon map on the following pages is a list of prominent features, including the days in the lunation when they are normally close to the terminator and thus easiest to see.

The dates of visibility vary slightly through the effects of **libration**. Because the Moon's orbit is inclined to the Earth's equator and also because it moves in an ellipse, the Moon appears to rock slightly from side to side (and nod up and down). Over time a total of 59% of the lunar terrain can be observed. Areas near the **limb** (the edge of the Moon) may vary considerably in their location and visibility. This is easily noticeable with Mare Crisium (Sea of Crises, eastern limb of the Moon, in the western sky) and the craters Tycho (southern part of the Moon) and Plato (northern part). Another effect is that at crescent phases before and after New Moon, the normally non-illuminated portion of the Moon receives a certain amount of light, reflected from the Earth. This **Earthshine** may enable certain bright features (such as the craters Aristarchus, Kepler and Copernicus on the western limb of the Moon, in the eastern sky) to be detected.

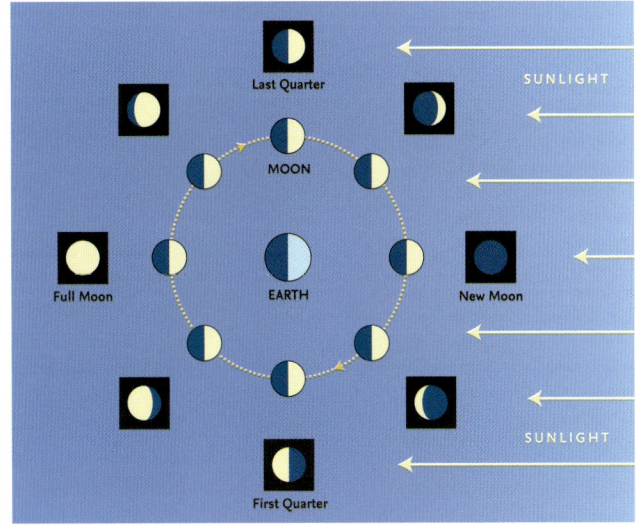

The Moon phases. *During its orbit around the Earth we see different portions of the illuminated side of the Moon's surface.*

Map of the Moon

Abulfeda	6:20	
Agrippa	7:21	
Albategnius	7:21	
Aliacensis	7:21	
Alphonsus	8:22	
Anaxagoras	9:23	
Anaximenes	11:25	
Archimedes	8:22	
Aristarchus	11:25	
Aristillus	7:21	
Aristoteles	6:20	
Arzachel	8:22	
Atlas	4:18	
Autolycus	7:21	
Barrow	7:21	
Billy	12:26	
Birt	8:22	
Blancanus	9:23	
Bullialdus	9:23	
Bürg	5:19	
Campanus	10:24	
Cassini	7:21	
Catharina	6:20	
Clavius	9:23	
Cleomedes	3:17	
Copernicus	9:23	
Cyrillus	6:20	
Delambre	6:20	
Deslandres	8:22	
Endymion	3:17	
Eratosthenes	8:22	
Eudoxus	6:20	
Fra Mauro	9:23	
Fracastorius	5:19	
Franklin	4:18	
Gassendi	11:25	
Geminus	3:17	
Goclenius	4:18	
Grimaldi	13-14:27-28	
Gutenberg	5:19	
Hercules	5:19	
Herodotus	11:25	
Hipparchus	7:21	
Hommel	5:19	
Humboldt	3:15	
Janssen	4:18	
Julius Caesar	6:20	
Kepler	10:24	
Landsberg	10:24	
Langrenus	3:17	
Letronne	11:25	
Linné	6	
Longomontanus	9:23	

The numbers indicate the day or days (the age) of the Moon when features are usually best visible.

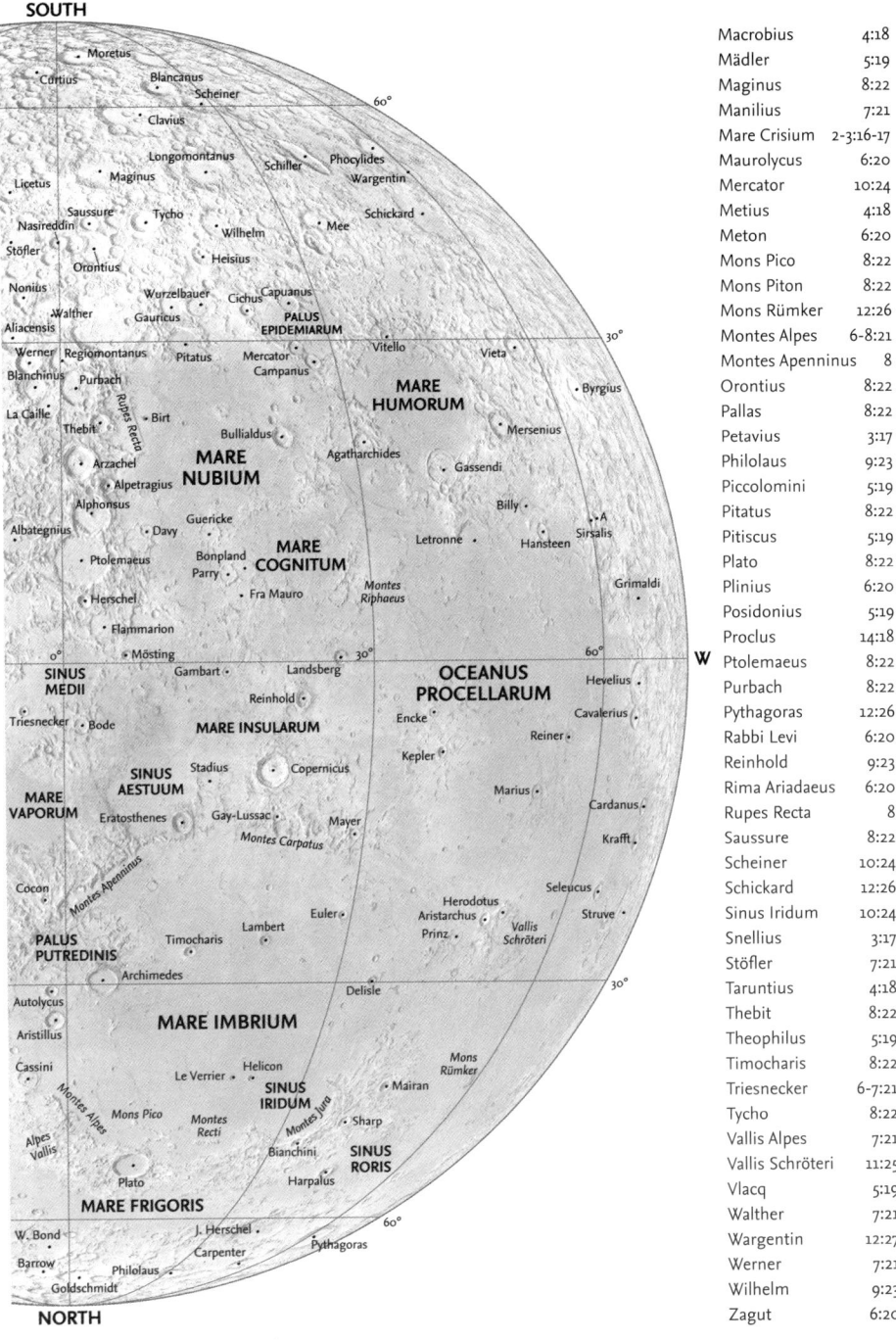

Feature	Time
Macrobius	4:18
Mädler	5:19
Maginus	8:22
Manilius	7:21
Mare Crisium	2-3:16-17
Maurolycus	6:20
Mercator	10:24
Metius	4:18
Meton	6:20
Mons Pico	8:22
Mons Piton	8:22
Mons Rümker	12:26
Montes Alpes	6-8:21
Montes Apenninus	8
Orontius	8:22
Pallas	8:22
Petavius	3:17
Philolaus	9:23
Piccolomini	5:19
Pitatus	8:22
Pitiscus	5:19
Plato	8:22
Plinius	6:20
Posidonius	5:19
Proclus	14:18
Ptolemaeus	8:22
Purbach	8:22
Pythagoras	12:26
Rabbi Levi	6:20
Reinhold	9:23
Rima Ariadaeus	6:20
Rupes Recta	8
Saussure	8:22
Scheiner	10:24
Schickard	12:26
Sinus Iridum	10:24
Snellius	3:17
Stöfler	7:21
Taruntius	4:18
Thebit	8:22
Theophilus	5:19
Timocharis	8:22
Triesnecker	6-7:21
Tycho	8:22
Vallis Alpes	7:21
Vallis Schröteri	11:25
Vlacq	5:19
Walther	7:21
Wargentin	12:27
Werner	7:21
Wilhelm	9:23
Zagut	6:20

MAP OF THE MOON

Eclipses in 2026

Lunar eclipses

There are two lunar eclipses in 2026. The first is a total eclipse, which takes place on 3 March and will be visible from eastern Asia, Australia as well as parts of North and South America. The second is a partial eclipse occurring on 28 August; this will be visible from Europe, Africa and the eastern Pacific. During a total lunar eclipse, the Moon moves eastward into the Earth's penumbra, then moves behind the Earth into the darkest region of the Earth's shadow – the umbra – eventually creeping back out into the penumbra and into the sunlight. On 3 March maximum eclipse occurs at 11:34 and the Moon will spend a total of 58 minutes immersed in the darkest shadow of the Earth. During this time the Moon is still visible; however, it will appear a reddish-brown colour. This is due to refracted sunlight which passes through the Earth's atmosphere and bends towards the lunar surface. Bluer hues are scattered outwards by the Earth's atmosphere, leaving behind redder light which eventually reaches the Moon. On the early morning of 28 August, the time of maximum eclipse will be 04:13; the umbral (partial) eclipse will last 3 hours 18 minutes.

Solar eclipses

There are two solar eclipses in 2026. The first is an annular eclipse that takes place on 17 February with the maximum occurring at 12:12, visible from Antarctica; it will appear as a partial eclipse from south Argentina, Chile and South Africa. The duration of the annular eclipse is 2 minutes 20 seconds. There will be a total eclipse on 12 August visible from Iceland, Spain, Greenland and the Arctic. Maximum occurs at 17:46, the duration is 2 minutes 18 seconds. A partial solar eclipse will be visible from Europe, west Africa and North America.

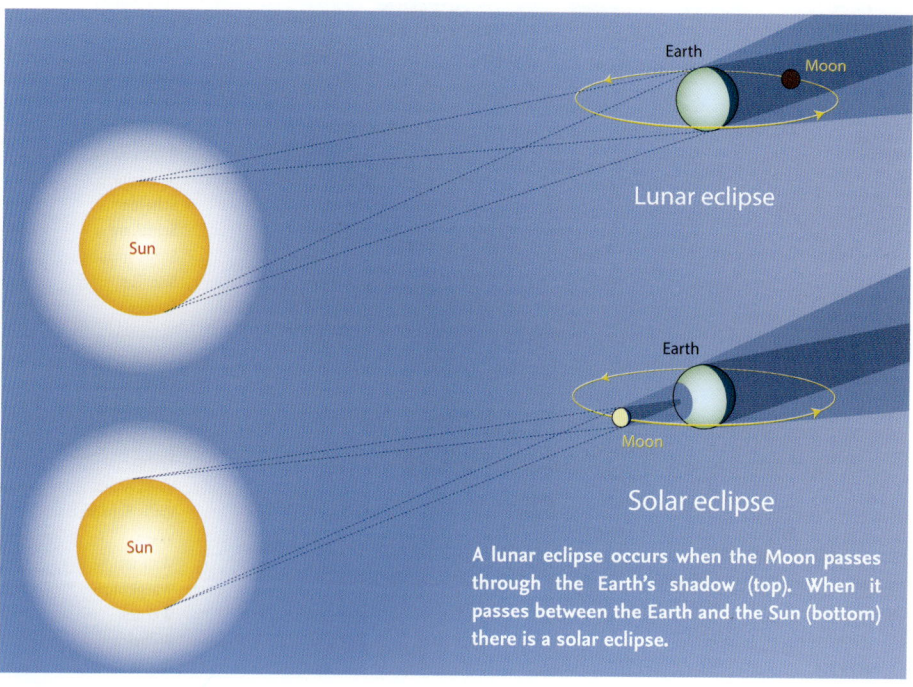

A lunar eclipse occurs when the Moon passes through the Earth's shadow (top). When it passes between the Earth and the Sun (bottom) there is a solar eclipse.

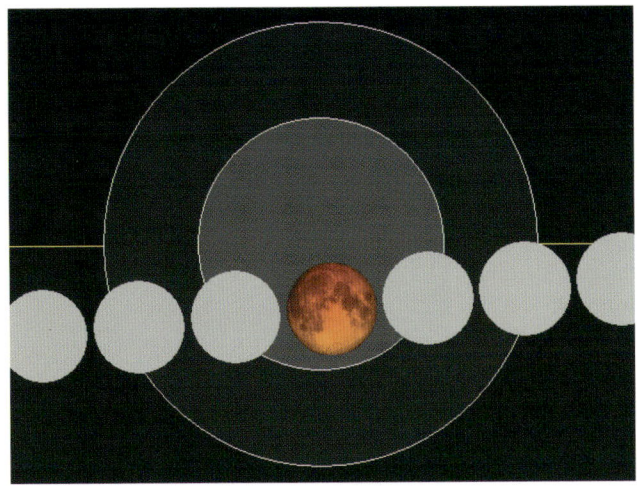

The path of the Moon as it passes through the shadow of the Earth on 3 March. Maximum eclipse occurs at 11:34 UT.

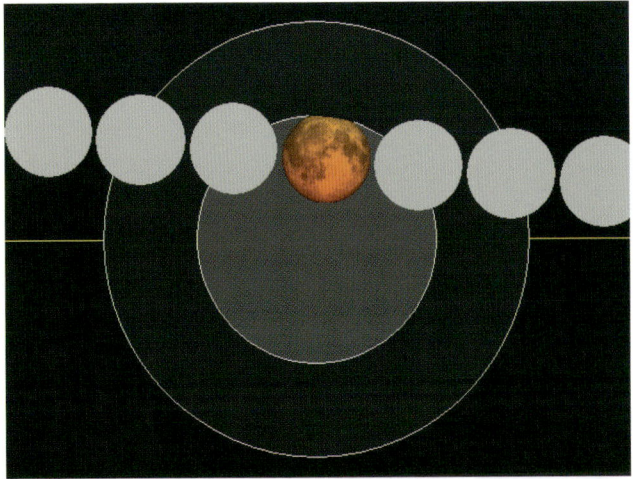

The path of the Moon as it passes through the shadow of the Earth on 28 August. Maximum eclipse occurs at 04:13 UT.

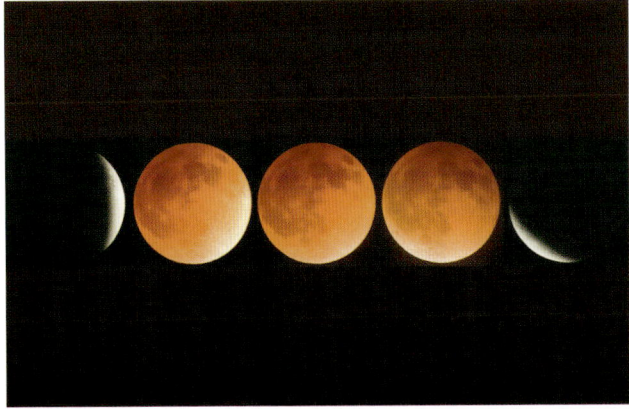

A sequence of five images showing the central stages of a lunar eclipse, taken from Cascade-Siskiyou National Monument, Oregon in 2014.

The Planets in 2026

Mercury and Venus

Mercury reaches greatest eastern elongation three times during 2026 – it will be seen in the early evening of 19 February (Venus and Saturn are close by), 15 June (Venus and Jupiter are setting with Mercury) and 12 October, although it will be just above the horizon at sunset on this day. Mercury switches to an appearance at dawn when it is at greatest western elongation on 3 April, 2 August and 20 November (with Venus close by), although in April it will be positioned at a low altitude with Mars adjacent to the east. Its magnitude oscillates throughout the year – it will be brightest in May (-2.4) and faintest in November (6.5).

Venus moves from the dawn to the evening sky at the beginning of the year. It reaches greatest eastern elongation on 15 August appearing as the evening star, lying in Virgo. Venus leads the Sun in the dawn sky in late October, moving apart to reach greatest western elongation on 3 January 2027. Venus is brightest (mag. -4.9) in late November and dimmest (mag. -3.9) throughout most of March.

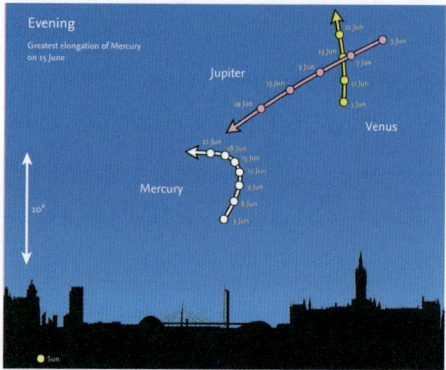

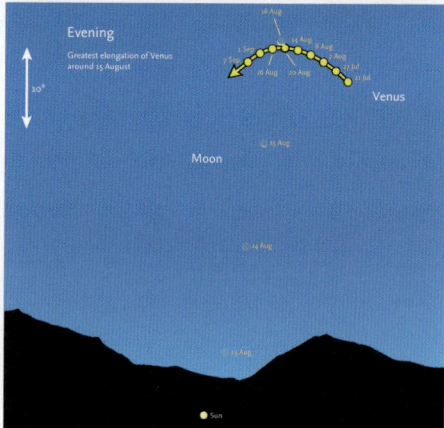

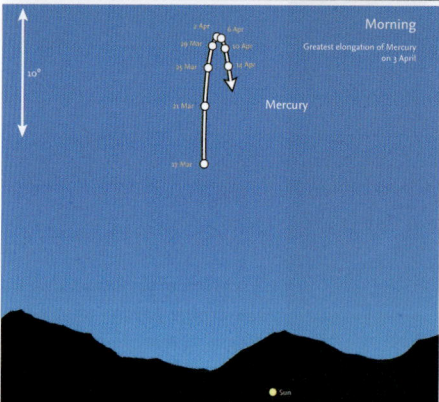

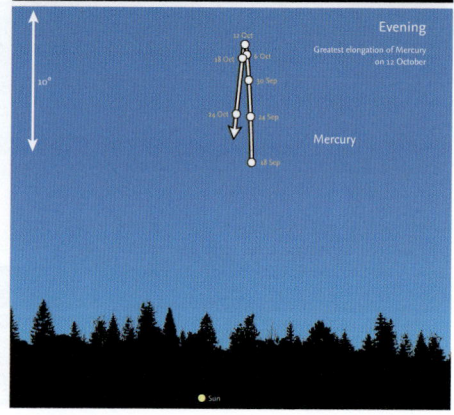

Mars

Mars is at its faintest in June (mag. 1.4) and it will brighten in the last half of the year, reaching a peak magnitude of -0.1 at the end of the year, when it will sit in **Leo**. Its apparent size in the sky will increase as it reaches opposition, which will next occur on 19 February 2027. Mars starts the year along the ecliptic in **Sagittarius**, moves into **Capricornus** in January and pulls away from the Sun in the sky, appearing at dawn by February. It progresses into **Aquarius**, **Pisces** in April and **Aries** in May. Mars will spend mid-summer in **Taurus**, eventually moving into **Gemini** in August, **Cancer** in early Autumn and it will stay in **Leo** from Halloween onwards. Mars will rise earlier, creeping above the horizon a few hours before midnight by the end of the year.

As it takes longer to complete its orbit than the Earth, Mars does not come to opposition every year; the last opposition took place 16 January 2025. Oppositions occur during a period of retrograde motion, when the planet appears to move westwards against the pattern of distant stars. Mars will continue to move eastward until 10 January 2027 when it will appear to move backwards as the Earth overtakes it in its orbit around the Sun.

Because of its eccentric orbit, which carries it at differing distances from the Sun (and Earth), not all oppositions of Mars are equally favourable for observation. The relative positions of Mars and the Earth are shown above. It will be seen that

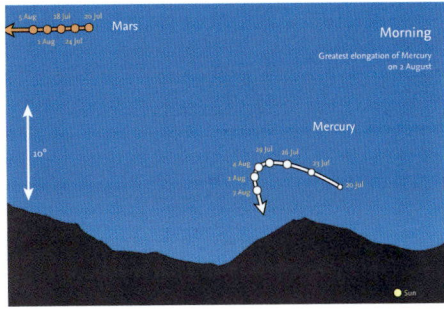

the opposition of 2018 was very close and thus favourable for observation, and that of 2020 was also reasonably good. By comparison, opposition in 2027 will be at a far greater distance, so the planet will not appear as large as it did in 2018.

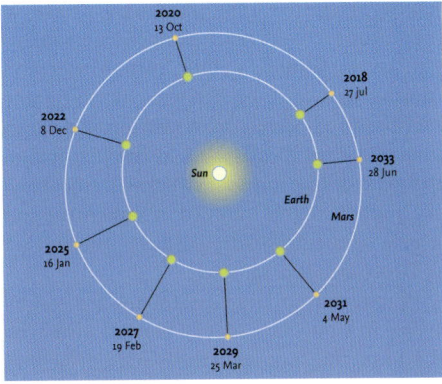

The oppositions of Mars between 2018 and 2033. The next opposition takes place 19 February 2027.

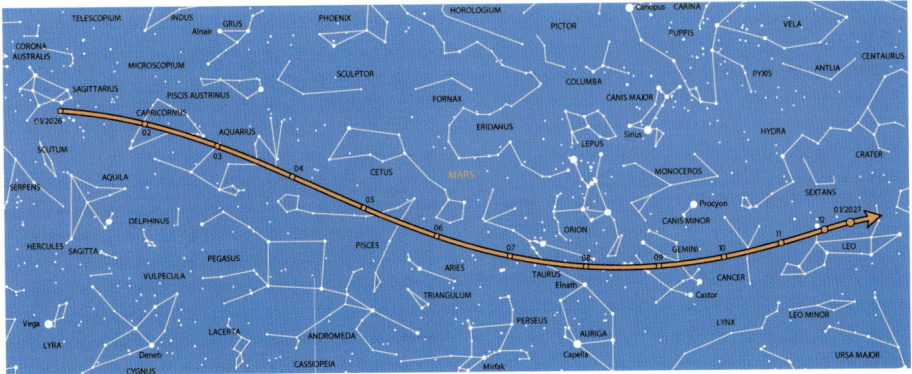

The path of Mars in 2026. Mars is visible at dawn for most of the year, eventually rising before midnight from November.

THE PLANETS

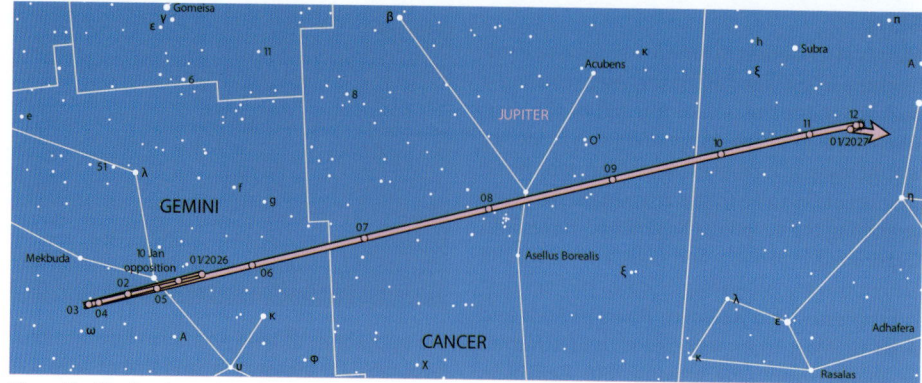

The path of Jupiter in 2026. Jupiter reaches opposition on 10 January. It ends retrograde motion on 11 March and re-enters westward motion on 13 December.

Jupiter and Saturn

Jupiter starts the year dazzlingly bright at magnitude -2.7. It reaches opposition on 10 January in **Gemini**, visible from sunset and moving westward. Jupiter ends its retrograde motion on 11 March. It moves into **Cancer** in June and dims to mag. -1.8 in late July, lost in the glare of the Sun as it switches to the dawn sky. Jupiter crosses into Leo in September and it re-enters retrograde westward motion on 13 December at mag. -2.3.

Jupiter's four largest satellites are readily visible in binoculars. Not all four are visible all the time, they are sometimes hidden behind the planet or invisible in front of it. Io, the closest to Jupiter, orbits in just under 1.8 days, and Callisto, the farthest away, takes about 16.7 days. In between are Europa (c. 3.6 days) and the largest, Ganymede (c. 7.2 days).

Saturn is in **Aquarius** at the start of the year, it moves into **Pisces** by the end of January, appearing shortly after sunset at its faintest (mag. 1.0). The ringed planet swings into the dawn sky in late March and pulls away from the Sun, creeping above the horizon before midnight from August. On 26 July Saturn will enter retrograde motion. It will reach opposition on 4 October and it will shine brightly with a magnitude of 0.3. This will be a good time to look at the planet although its rings will be close to edge-on, inclined by only 7° to our line of sight (allowing observers to see more of the southern side). Saturn will end its retrograde movement on 11 December.

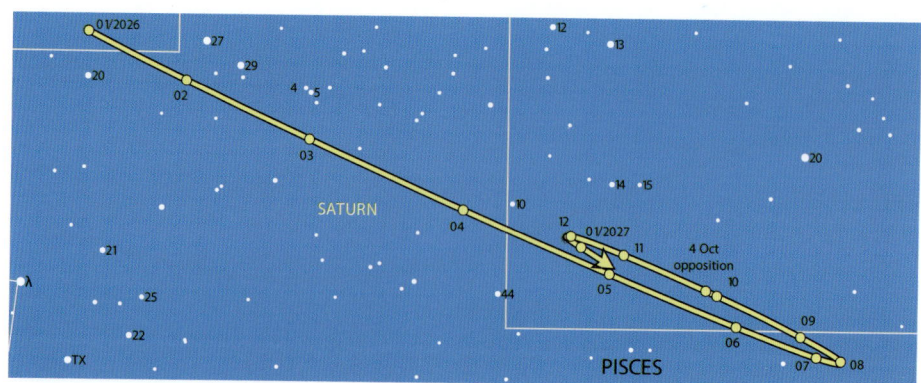

The path of Saturn in 2026. Saturn enters retrograde motion on 26 July, reverting back to direct motion on 11 December. It reaches opposition on 4 October.

Uranus and Neptune

Uranus sits in Taurus for the whole of 2026. It leads the Sun in the dawn sky in late May. It ends its retrograde motion on 4 February (it has been moving westward since September 2025). It will re-enter retrograde motion on 10 September, moving westward across the sky until 8 February 2027. It will reach opposition on 25 November (mag. 5.6).

Neptune moved into **Pisces** in May 2022; it will stay in Pisces until August 2039. In March it crosses from the evening to the dawn sky, sitting low in the east before sunrise for the rest of Spring. It enters retrograde motion on 7 July at mag. 7.9, visible after midnight. It will reach opposition on 26 September (mag. 7.7). It reverts to direct motion on 12 December and will continue moving eastwards until 10 July 2027.

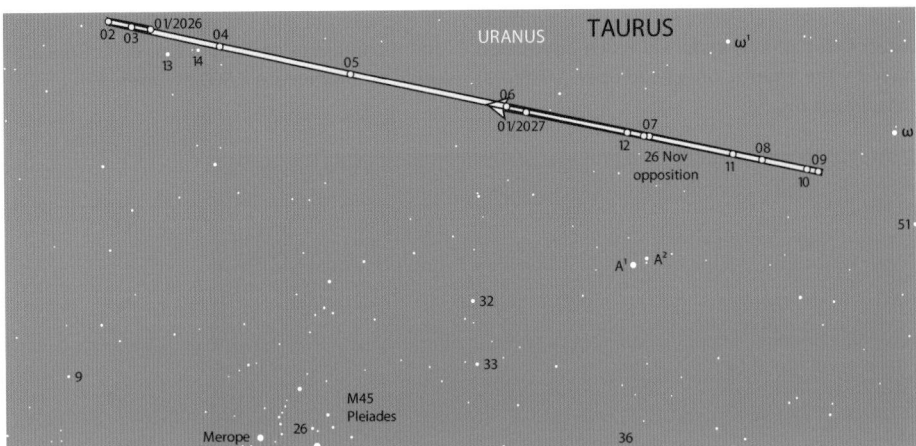

The path of Uranus in 2026 in Taurus. Uranus ends retrograde motion on 4 February, re-entering on 10 September. It reaches opposition on 25 November.

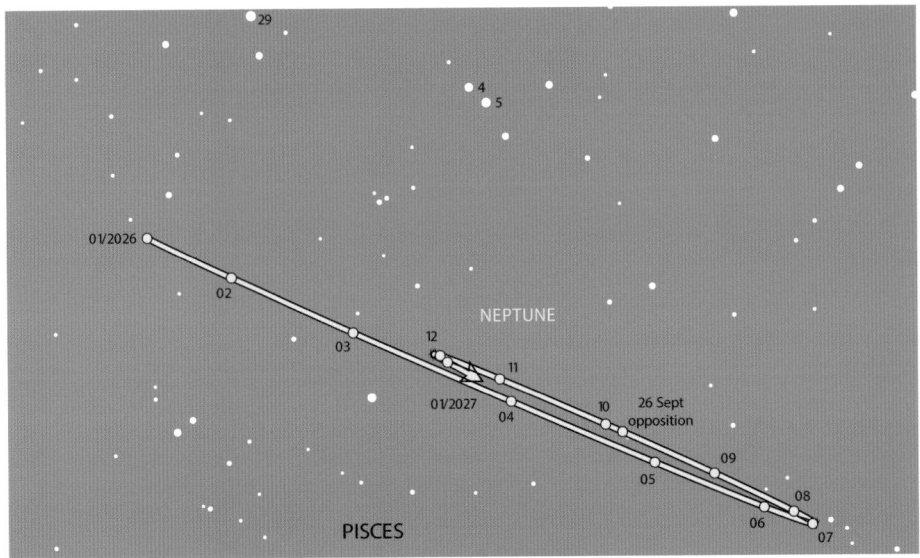

The path of Neptune in 2026 in Pisces. Pisces enters retrograde motion on 7 July, reaching opposition on 26 September. It reverts to direct eastward motion on 12 December.

Minor Planets in 2026

Three minor planets come to opposition and rise above magnitude 9.0 in 2026. They are closest to Earth at this point and reach their highest point in the sky around midnight – this is the best time to see them. On 30 September the asteroid **(192) Nausikaa** will lie in **Pisces**, reaching a peak magnitude of 8.4 at opposition. On 4 October **(2) Pallas** will reach magnitude 8.2 in **Cetus** and **(4) Vesta** will reach opposition later in the month on 13 October, visible at magnitude 6.5.

In December the dwarf planet **(1) Ceres** will increase in brightness from magnitude 7.6 to 6.9, reaching opposition on 7 January 2027 at a distance of 1.63 AU (Astronomical Units, equivalent to the average distance from the Earth to the Sun). Ceres is the only dwarf planet that lies within the orbit of Neptune (it is part of the asteroid belt between Mars and Jupiter) and it completes its orbit in 4.6 years. The Earth could accommodate 2,500 of Ceres.

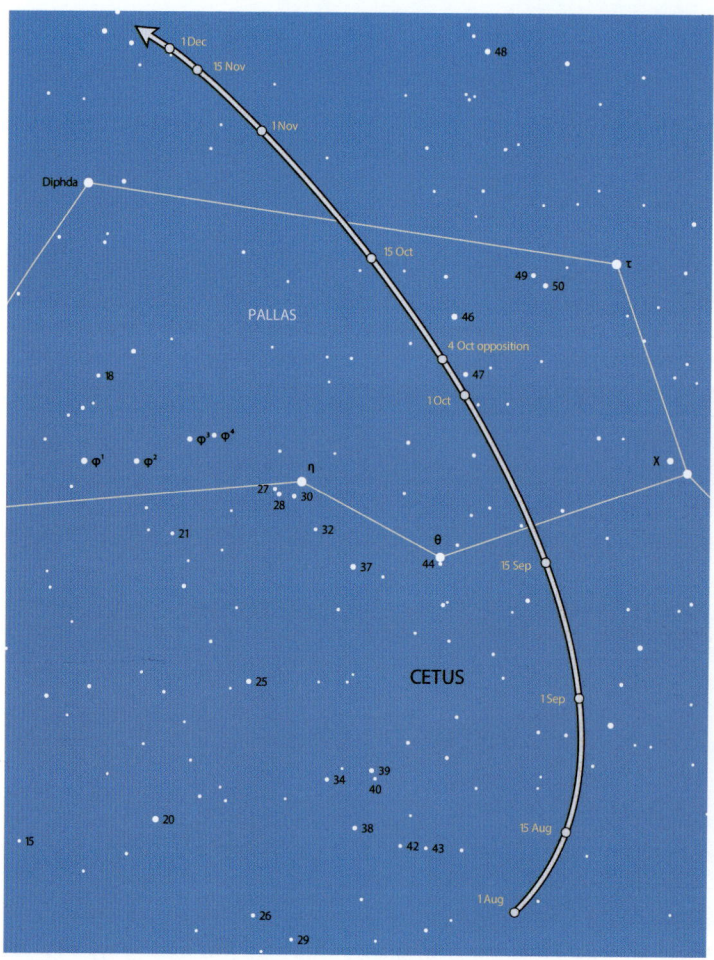

The path of (2) Pallas in 2026. (2) Pallas reaches opposition on 4 October 2026.

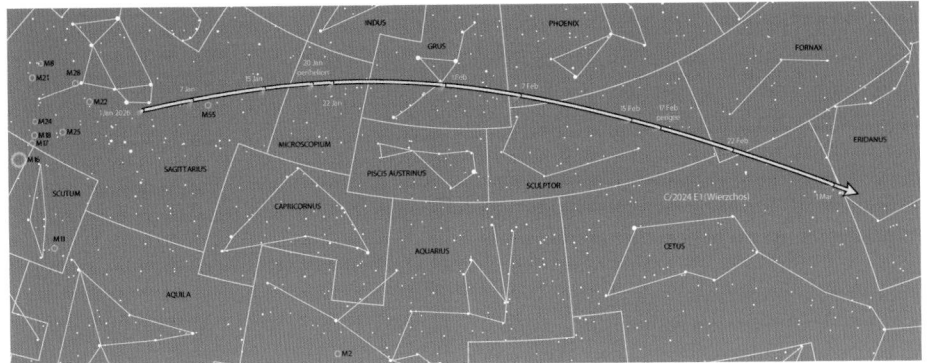

Comet C/2024 E1 (Wierzchos) reaches magnitude 6 at its closest approach to Earth on 17 February.

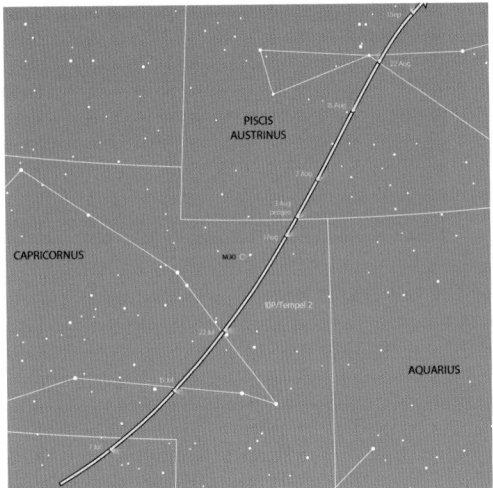

Comet 10P/Tempel 2 reaches magnitude 7 at its closest approach to Earth on 3 August.

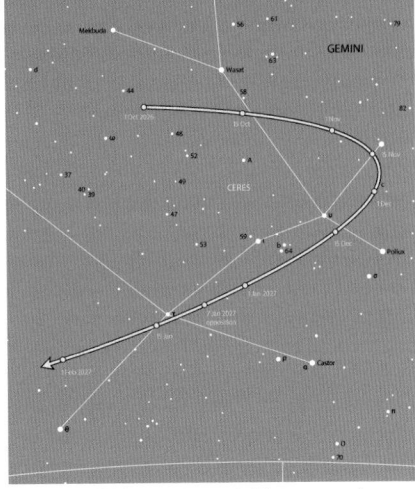

The path of (1) Ceres in 2026. (1) Ceres reaches opposition on 7 January 2027.

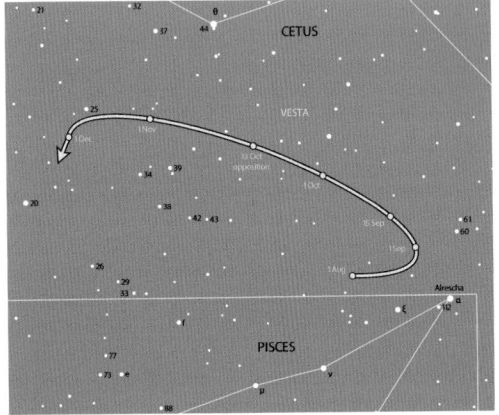

The path of (4) Vesta in 2026. (4) Vesta reaches opposition on 13 October.

MINOR PLANETS

Comets in 2026

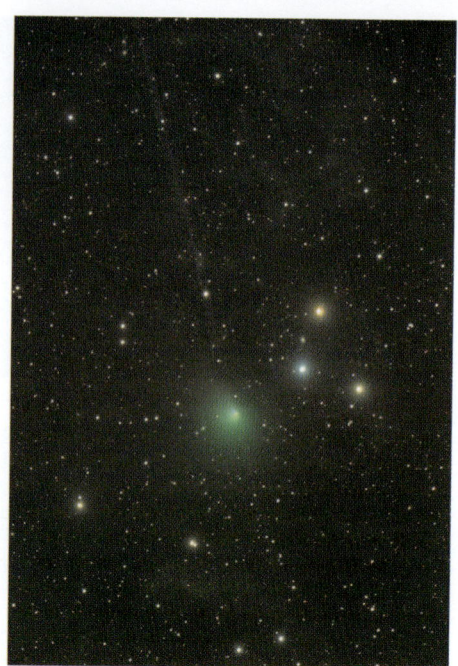

Although comets may occasionally become very striking objects in the sky, their occurrence and particularly the existence or length of any tail and their overall magnitude are notoriously difficult to predict. Naturally, it is only possible to predict the return of periodic comets (whose names have the prefix 'P'). Many comets appear unexpectedly (these have names with the prefix 'C'). Bright, readily visible comets such as C/1995 Y1 Hyakutake, C/1995 O1 Hale-Bopp, C/2006 P1 McNaught or C/2020 F3 NEOWISE are rare. Most periodic comets are faint and only a very small number ever become bright enough to be easily visible with the naked eye or with binoculars.

Comet 24P/Schaumasse will be visible most of the night from January to March, brightening from magnitude 9 to 8, moving through the constellations of Virgo, Boötes and Serpens Caput. It will make its closest approach on 4 January 2026, passing within a distance equivalent to 60% of the Earth–Sun distance. Comet C/2024 E1 (Wierzchos) will pop into Cetus, Eridanus and Taurus in early Spring, brightening to magnitude 6 at its closest point to Earth on 17 February. Comet 10P/Tempel 2 appears in June in Aquila at mag. 9, visible after 23:00. It moves into Capricornus in July and August, brightening to mag. 7 at point of closest approach on 3 August. Comet 10P/Tempel 2 has a period of 5.4 years.

Comet 10P/Tempel 2 showing the coma (halo) of gas surrounding the nucleus.

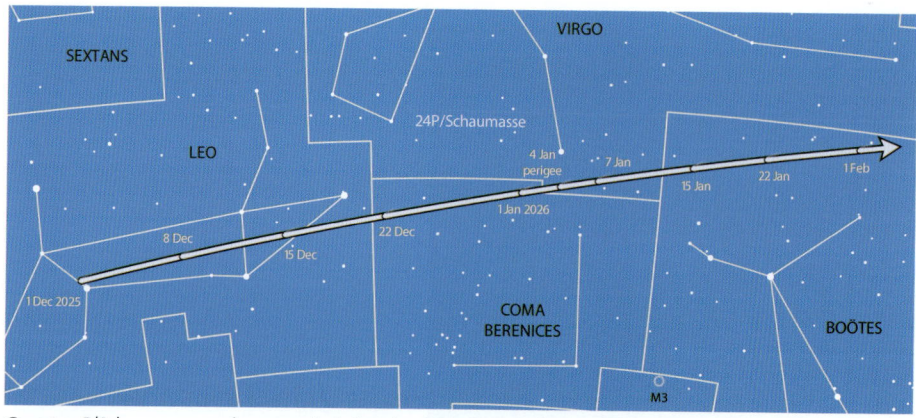

Comet 24P/Schaumasse reaches magnitude 8 at its closest approach to Earth on 4 January.

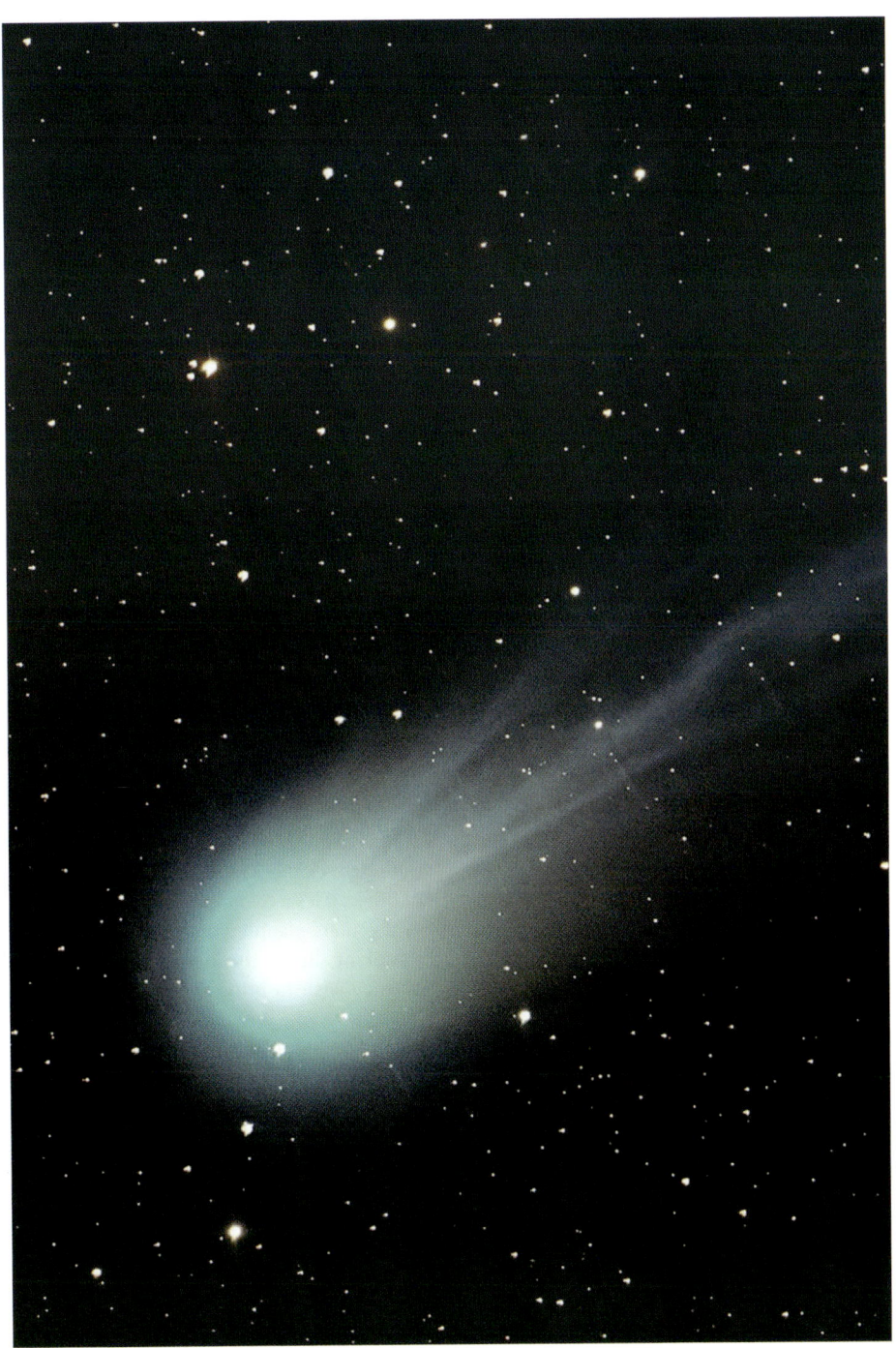
Comet C/2020 F8 SWAN photographed from Indonesia by Christian Gloor.

Introduction to the Month-by-Month Guide

The monthly charts

The pages devoted to each month contain a pair of charts showing the appearance of the night sky, looking north and looking south. The charts are drawn for the latitude of 35°S, so observers farther south will see slightly more of the sky on the southern horizon, and slightly less on the northern. Stars close to the horizon are always dimmed by atmospheric absorption (this is called extinction), so sometimes the faintest stars marked on the charts may not be visible.

The charts are drawn to show the appearance at 23:00 UT for the 1st of each month. The same appearance will apply an hour earlier (22:00 UT) on the 15th, and yet another hour earlier (21:00 UT) at the end of the month (shown as the 1st of the following month). Daylight Saving Time (DST) is not used in South Africa. In New Zealand, the previous period of DST ends on the first Sunday of April (5 April 2026) and starts on the last Sunday of September (27 September 2026). In those Australian states with DST, it ends on the first Sunday in April (5 April 2026) and starts on the first Sunday in October (4 October 2026). The appropriate times are shown on the monthly charts. Times of specific events are shown in the 24-hour clock of Universal Time (UT) used by astronomers worldwide and the correct zone time (and DST, where appropriate) may be found from the details inside the front cover.

The charts may be used for earlier or later times during the night. To observe two hours earlier, use the charts for the preceding month; for two hours later, the charts for the next month.

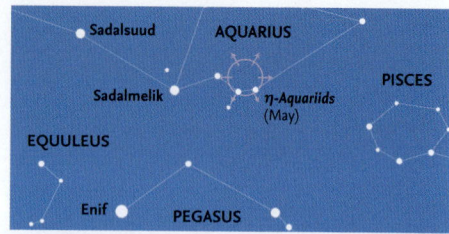

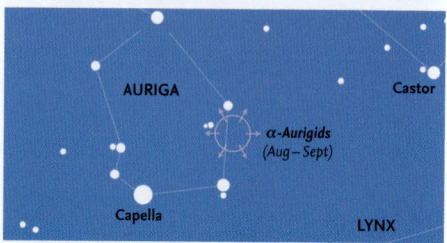

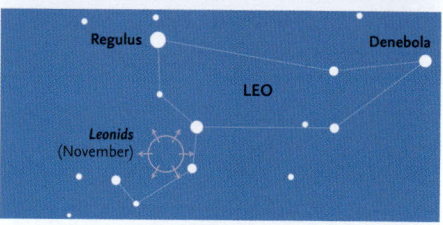

Tips for observing the night sky

A successful observing session requires planning, flexibility and patience. Check the weather forecast, sunset and moonrise and moonset times and choose a location where you have a clear view of the horizon wherever possible. Bring your star guide or app on a mobile device, a red torch to aid your night vision and a pair of binoculars and a tripod to keep them steady. You may be outside for a while – it's advised that you wear warm clothes, socks and shoes and bring hot drinks if it's a cold night.

If you are planning on looking for meteors or looking up at the night sky for a while you may wish to bring a deckchair or picnic blanket with you. When looking at faint objects (unlike the Moon) you will need to give your eyes 10–20 minutes to adapt to the dark and achieve night vision. If you are using your phone, switch it to night vision mode and use a red torch to find your way around so that you maintain your night vision.

Astrophotography

You can capture the Moon and five planets visible to the naked eye or the brightest stars using your smartphone. The challenge is keeping your phone steady and gaining control

Shower	Dates of activity 2026	Date of maximum 2026	Possible hourly rate
α-Centaurids	28 January to 21 February	8 February	6
γ-Normids	25 February to 28 March	14 March	6
π-Puppids	15 April to 28 April	24 April	var.
η-Aquariids	19 April to 28 May	6 May	40
α-Capricornids	3 July to 15 August	30 July	5
Southern δ-Aquariids	12 July to 23 August	30 July	25
Piscis Austrinids	15 July to 10 August	28 July	5
Perseids	17 July to 24 August	13 August	150
α-Aurigids	28 August to 5 September	1 September	6
Southern Taurids	10 September to 20 November	10 October	5
Orionids	2 October to 7 November	21 October	15
Northern Taurids	20 October to 10 December	12 November	5
Leonids	6 November to 30 November	18 November	15
Phoenicids	28 November to 9 December	2 December	var.
Puppid-Velids	1 December to 15 December	7 December	10
Geminids	4 December to 20 December	14 December	120

over the focus and exposure (if the camera allows it). It is advised to attach the phone to a tripod, or you could use an adaptor to attach your phone to the eyepiece of a telescope – this will allow you to align it correctly with minimal effort. You can use the zoom function on your camera to change the magnification of the image.

There are many apps available that will provide more control of your camera settings and allow you to take better quality images: NightCap Pro, ProCam and AstroShader for iOS and Open Camera, DeepSkyCamera or Camera FV-5 for Android. For the Moon and bright planets, your camera should auto lock and focus, otherwise manually set the focus to infinity and keep the exposure short (a fraction of a second if it is a gibbous or Full Moon, one or two seconds for a thin crescent Moon and planets). You could control the shutter remotely via the volume control on wireless headphones, or set a timer on your phone (3 seconds delay). For fainter objects, increase the exposure and the ISO (the sensitivity of the camera detector); however, a higher ISO setting will introduce more noise. Trial and error will help you choose the best settings. Some apps allow images to be stacked; this means you can overlay images to increase brightness further. This is useful for the constellations, clusters such as the Pleiades and nebulous objects such as the Orion Nebula and Andromeda.

Meteors

Details of specific meteor showers are given in the months when they come to maximum. Radiants that lie below the horizon or in constellations not readily visible in that month are not marked. For this reason, special charts for the Eta (η) Aquariids (May), the Alpha (α) Aurigids (August and September) and the Leonids (November) are given here. Meteors from such showers may still be seen, because the most effective region for seeing meteors is some 40–45° from the radiant, and that area of sky may well be above the horizon. A table of the best meteor showers visible during the year is also given here. The rates given are those that an experienced observer might see under ideal conditions. Generally, the observed rates will be far lower.

The photographs

One or more photographs of constellations visible in certain specific months are included. It should be noted, however, that several factors affect observation: differences between the sensitivity of different individuals to faint starlight including colour, the degree to which they have become adapted to the dark, the local altitude of the star and the atmospheric conditions. Also, the exposure time can be altered on a digital camera, thus enhancing the intensity and colour of the object.

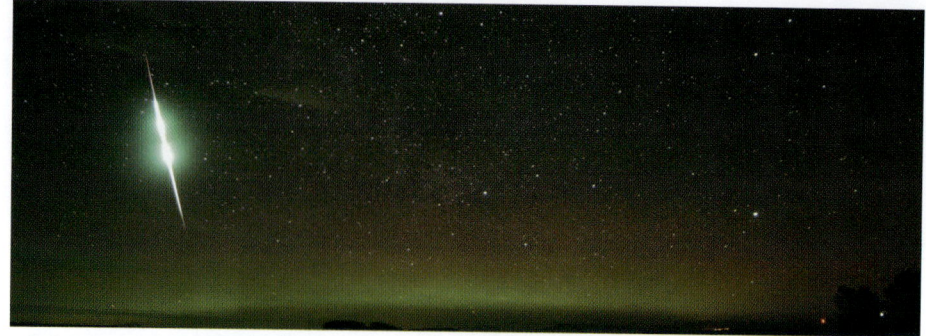

A fireball (with flares approximately as bright as the Full Moon), photographed against a weak auroral display from Tarbat Ness by Denis Buczynski in Scotland on 22 January 2017.

The Moon calendar

The Moon calendar shows the phase of the Moon for every day of the month. Also shown is the **age** of the Moon (the day in the **lunation**), beginning at New Moon, which may be used to determine the best time for observation of specific lunar features.

The Moon

The section on the Moon includes details of any lunar or solar eclipses that may occur during the month. Similar information is given about any important occultations. Mainly, however, this section summarizes when the Moon passes close to planets or the five prominent stars close to the ecliptic. The dates when the Moon is closest to the Earth (at **perigee**) and farthest from it (at **apogee**) are shown in the monthly calendars, and only the nearest and farthest points during the year are mentioned.

The planets and minor planets

Brief details are given of the location, movement and brightness of the planets from Mercury to Saturn throughout the month. Mercury and Venus are normally easiest to see around either eastern or western elongation, in the evening or morning sky, respectively.

The dates at which the superior planets reverse their motion (from direct motion to **retrograde**, and retrograde to direct) and of opposition (when a planet generally reaches its maximum brightness) are given. Jupiter and Saturn are normally easiest to see around opposition, which occurs every year. Mars, by contrast, moves relatively rapidly against the background stars and in some years never comes to opposition. The distant outer planets, especially Saturn, Uranus and Neptune, may spend most or all of the year in a single constellation.

Uranus is normally magnitude 5.7 to 5.9, and thus at the limit of naked-eye visibility under exceptionally dark skies, but bright enough to be readily visible in binoculars. Similar considerations apply to Neptune, it is always fainter (generally magnitude 7.8 to 8.0); however, it is still visible in most binoculars. The charts on page 25 show their positions during 2026.

In any year, few minor planets ever become bright enough to be detectable in binoculars. Just one, (4) Vesta, on rare occasions brightens sufficiently for it to be visible to the naked eye. Our limit for visibility is magnitude 9.0 and details and charts are given for those objects that exceed that magnitude during the year, normally around opposition. Minor planet charts for 2026 are on pages 26 and 27.

The ecliptic charts

Although the ecliptic charts are primarily designed to show the positions and motions of the major planets, they also show the motion of the Sun during the month. The light-tinted area shows the area of the sky that is invisible during daylight, but the darker area gives an indication of which constellations are likely to be visible at some time of the night.

The monthly calendar
For each month, a calendar shows details of significant events, including when planets are close to one another in the sky, close to the Moon, or close to any one of five bright stars that are spaced along the ecliptic. The times shown are given in Universal Time (UT).

The diagrams of interesting events
Each month, a number of diagrams show the appearance of the sky when certain events take place. However, the exact positions of celestial objects and their separations greatly depend on the observer's position on Earth. When the Moon is one of the objects involved, because it is relatively close to Earth, there may be very significant changes from one location to another. Close approaches between planets or between a planet and a star are less affected by changes of location, which may thus be ignored.

The diagrams showing the appearance of the sky are drawn for the latitude 35°S and longitude 150°E (approximately that of Sydney, Australia), so they will be approximately correct for much of Australia. However, for an observer farther north (say, in Brisbane or Darwin), a planet or star listed as being north of the Moon will appear even farther north, whereas one south of the Moon will appear closer to it – or may even be hidden (occulted) by it. For an observer at a latitude greater than 35°S (such as in New Zealand), there will be corresponding changes in the opposite direction: for a star or planet south of the Moon, the separation will increase, and for one north of the Moon, the separation will decrease. This is particularly important when occultations occur, which may be visible from one location, but not another.

The details given regarding the positions of the various bodies should be used as a guide to their location. A similar situation arises with the times that are shown. Similarly, dates and times are given, even if they fall in daylight, when the objects are likely to be completely invisible. However, such times do give an indication that the objects concerned will be in the same general area of the sky during both the preceding and the following nights.

Data used in this Guide
The data given in this Guide, such as timings and distances between objects, have been determined by the following sources: the Astronomical Applications Department of the US Naval Observatory; the Sky Events Calendar by Fred Espenak and Sumit Dutta (NASA's Goddard Space Flight Center); the NASA Jet Propulsion Laboratory (JPL) Horizons System; the International Astronomical Union's (IAU) Minor Planet Center; In-The-Sky.org.

Key to the symbols used on the monthy star maps.

STAR MAGNITUDES								PLANETS
● -1	● 0	● 1	● 2	● 3	· 4	Fainter	Milky Way	○ MERCURY
								○ VENUS
OBJECTS								○ MARS
○ Open Star Cluster		✧ Planetary Nebula				Meteor Radiant		○ JUPITER
◎ Globular Star Cluster		▭ Bright Nebula				Galaxy		○ SATURN

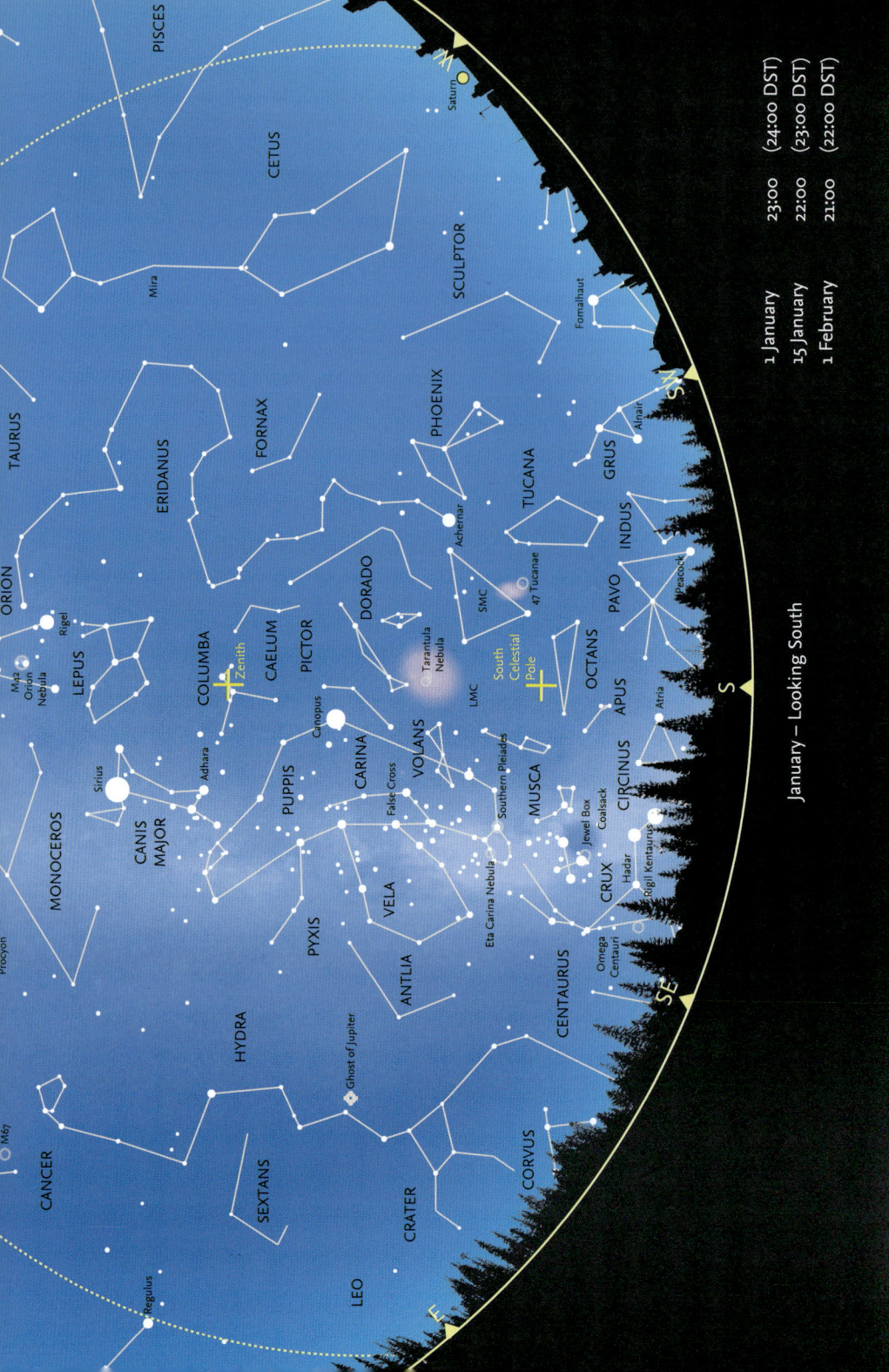

January – Looking South

Crux is quite prominent as it rises in the east, and even **Rigil Kentaurus** and **Hadar** (α and β Centauri), although low, are becoming easier to see. The whole of **Carina** is visible, with both the **Eta Carinae Nebula** and the **Southern Pleiades** between Crux and the **False Cross** of stars from the constellations of Carina and **Vela**. **Canopus** (α Carinae) is high in the south, roughly three-quarters of the way from the horizon to the zenith. Also in the south, but slightly lower, is the **Large Magellanic Cloud** (LMC), a satellite dwarf galaxy 160,000 light-years away, consisting of an estimated 20 billion stars. **Achernar** (α Eridani) is prominent between the constellations of **Hydrus** and **Phoenix**. The **Small Magellanic Cloud** (SMC), another satellite galaxy 200,000 light-years away and **47 Tucanae**, a globular cluster of around 10,000 stars are well-placed for observation alongside Hydrus. **Fomalhaut** (α Piscis Austrini) may be glimpsed low on the horizon towards the west, as may **Peacock** (α Pavonis) farther towards the south. The whole of **Pavo** and **Grus** is visible, together with most of **Indus**.

Meteors

There is one minor shower at the start of the year for observers in the southern hemisphere, the α-Centaurids (active late January to February), which has a low maximum rate of around six per hour.

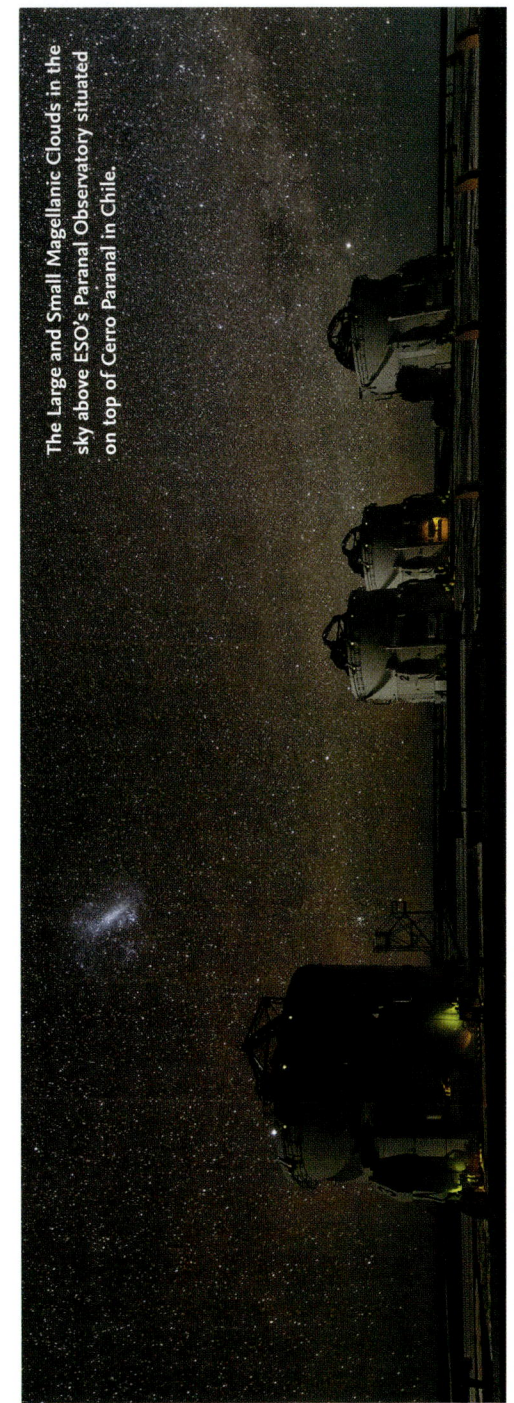

The Large and Small Magellanic Clouds in the sky above ESO's Paranal Observatory situated on top of Cerro Paranal in Chile.

JANUARY 35

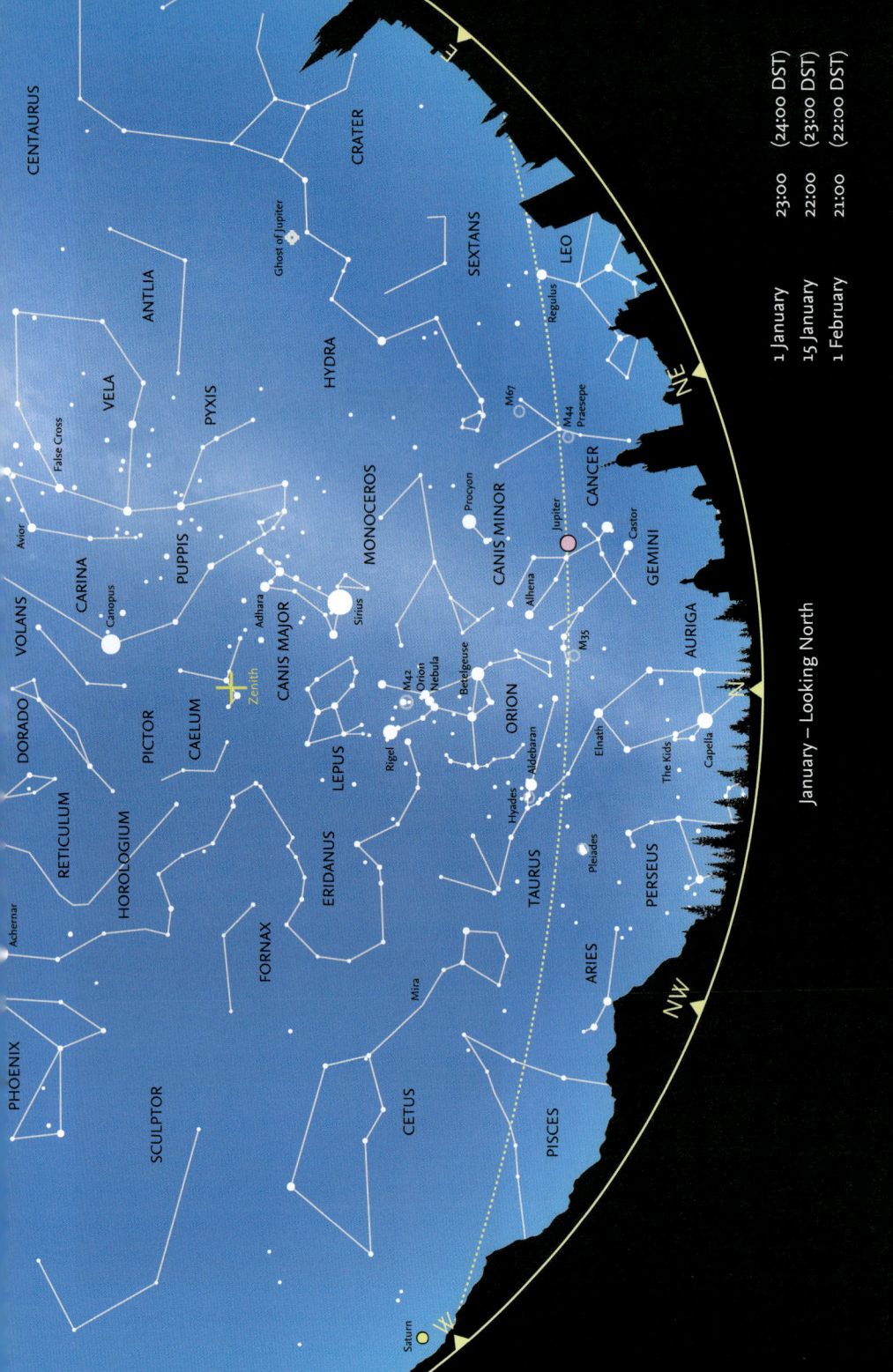

January – Looking North

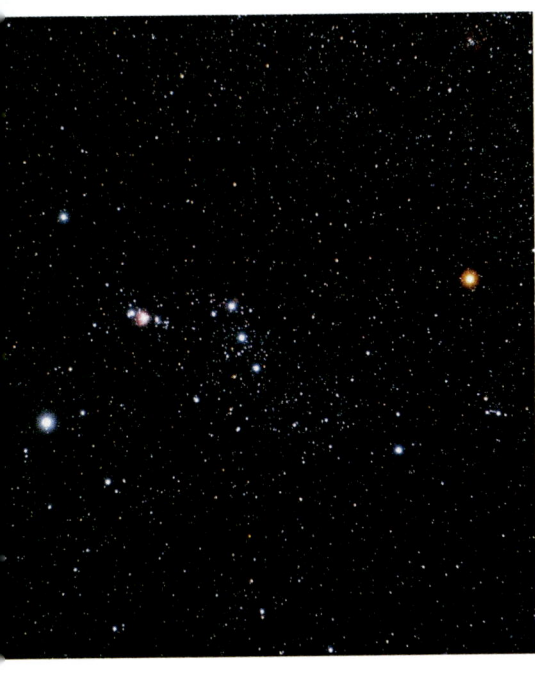

The northern sky is dominated by **Orion**, prominent during the summer months and visible at some time during the night. It is highly distinctive, with a line of three stars that form Orion's Belt. To most observers, the bright star **Betelgeuse** (α Orionis) shows a reddish tinge; it is known as a red supergiant star, in contrast to the brilliant bluish-white **Rigel** (β Orionis). The three stars of the belt lie directly south of the celestial equator. A vertical line of three 'stars' forms the 'sword' that hangs south of the 'belt'. With good viewing, the central 'star' appears as a hazy spot, even to the naked eye, and is actually the **Orion Nebula**, a star formation region. Binoculars reveal the four stars of the **Trapezium** cluster, which illuminate the nebula.

The Belt points down to the northwest towards **Taurus** (the Bull) and orange-tinted **Aldebaran** (α Tauri). Close to Aldebaran, there is a conspicuous 'V' of stars, called the **Hyades** cluster. Aldebaran lies much closer to the Earth than the Hyades, it just so happens to share the same line of sight. Farther along, the same line from Orion passes above a bright cluster of stars, the **Pleiades**, or Seven Sisters. Even the smallest pair of binoculars reveals this as a beautiful group of bluish-white stars. Another conspicuous star in Taurus, **Elnath** (β Tauri), lies directly above Orion and forms a big 'V' with Aldebaran.

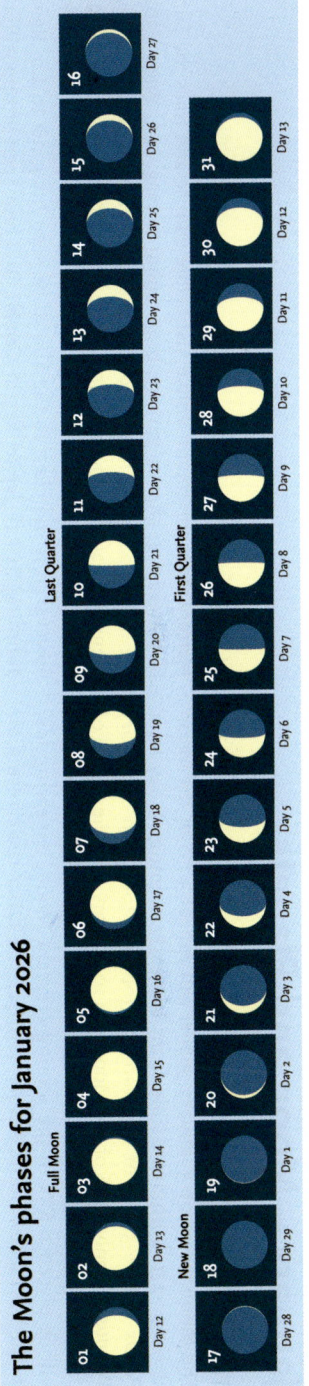

The constellation of Orion dominates the southern part of the sky during this period of the year; the three stars marking the belt are easy to find. Here, orange Betelgeuse, blue-white Rigel and the pink Orion Nebula are prominent.

The Moon's phases for January 2026

JANUARY 37

January – Moon and Planets

The Moon

The Moon is Full on 3 January, **Jupiter** will appear 3.7°S of the Moon at mag. -2.7, both are visible from the early evening. A day later **Pollux** will be 3.0°N of the waning gibbous Moon. On 6 January **Regulus** lies only 0.5°S of the Moon. On 10 January **Spica** lies 1.6°N of the Last Quarter Moon; four days later the red-orange star **Antares** will be placed 0.6°N of the waning crescent Moon, appearing low in the east a few hours before sunrise. The lunar cycle resets on 18 January with a New Moon. On 23 January **Saturn** (mag. 1.0) will appear 4.3°S of the waxing crescent Moon in the early evening, both setting shortly after 20:30. Three days later the Moon will be First Quarter, a day later the **Pleiades** cluster lies 1.1°S of the Moon. The month ends with **Jupiter** (mag. -2.6) 3.8°S of the waxing gibbous Moon and **Pollux** lies 3.0°N.

The Planets

Mercury starts the month visible just before sunrise and trails the Sun in the evening after 21 January, brightening to mag. -1.3. **Venus** leads the Sun at dawn at the start of the year and transitions to the early evening sky after 6 January, visible at the end of the month at mag. -3.9. **Mars** approaches the Sun in Sagittarius and swings into the dawn sky; however, it is too close to the Sun to be seen (mag. 1.1 to 1.0, then dimming to 1.2). **Jupiter** lies in **Gemini**, it is best seen at opposition at midnight on 10 January, mag. -2.7. **Saturn** moves from **Aquarius** to **Pisces** during the month (mag. 1.0). **Neptune** is nearby. **Uranus** sits in **Taurus** until a few hours after midnight, at mag. 5.6. **Neptune** is in **Pisces** at mag. 7.8.

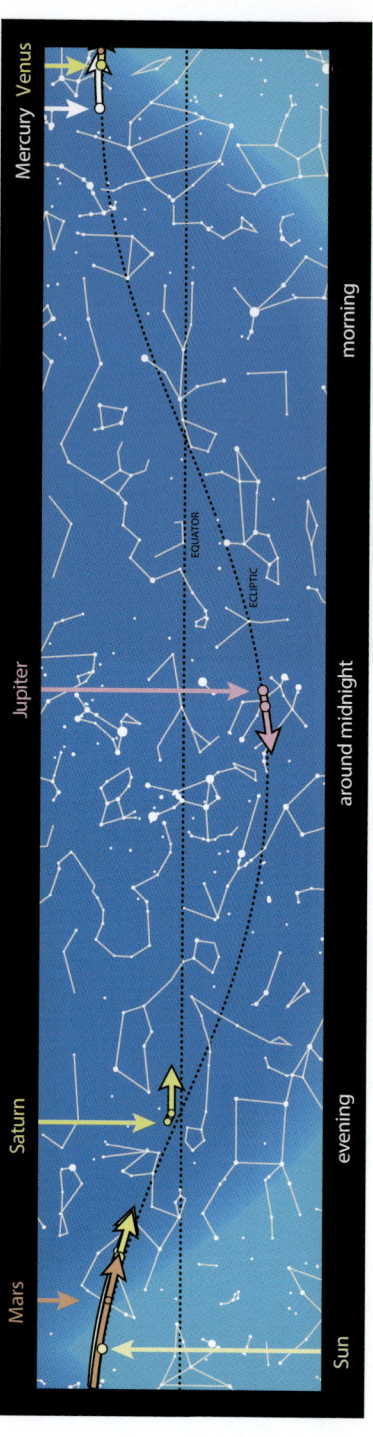

The path of the Sun and the planets along the ecliptic in January.

Calendar for January

01	21:43	Moon at perigee = 360,348 km
03	10:03	Full Moon
03	17:15	Earth at perihelion (147,099,900 km = 0.9833 AU)
03	22:01	Jupiter (mag. -2.7) 3.7°S of the Moon
04	03:28	Pollux 3.0°N of the Moon
06	16:20	Regulus 0.5°S of the Moon. An occultation will be visible from Mexico, Australia & New Zealand
10	08:34	Jupiter (mag. -2.7) at opposition
10	15:48	Last Quarter Moon
10	23:50	Spica 1.6°N of the Moon
13	20:48	Moon at apogee = 405,437 km
14	19:28	Antares 0.6°N of the Moon. An occultation will be visible from Australia
18	19:52	New Moon
23	12:31	Saturn (mag. 1.0) 4.3°S of the Moon
26	04:47	First Quarter Moon
27	21:07	Pleiades 1.1°S of the Moon
29	21:53	Moon at perigee = 365,878 km
31	02:31	Jupiter (mag. -2.6) 3.8°S of the Moon
31	13:45	Pollux 3.0°N of the Moon

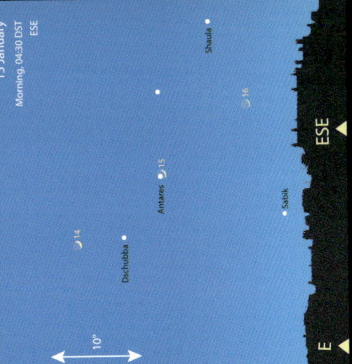

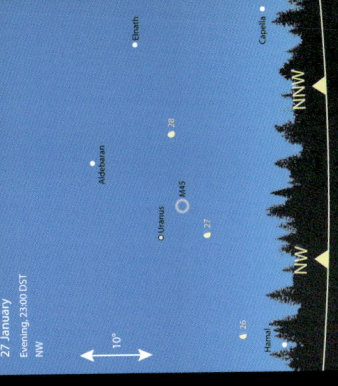

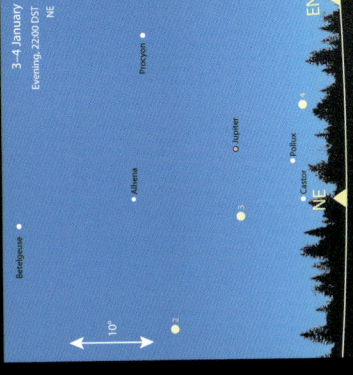

3–4 January • *The Full Moon lies in Gemini close to Jupiter (mag. -2.7). Pollux and Castor can be seen below the Moon, Procyon lies eastward.*

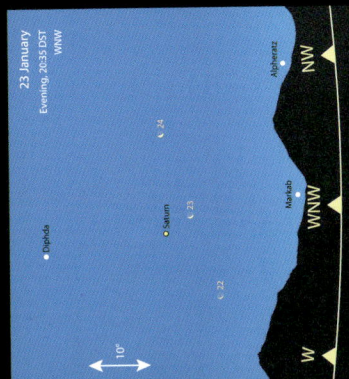

15 January • *Red-orange Antares sits next to the waning crescent Moon in the dawn sky, low in the southeast.*

23 January • *The waxing crescent Moon lies close to Saturn. Diphda (β Cet) is close by.*

27 January • *The waxing gibbous Moon is next to the Pleiades and Uranus, with Aldebaran, Capella and Elnath to the east.*

JANUARY 39

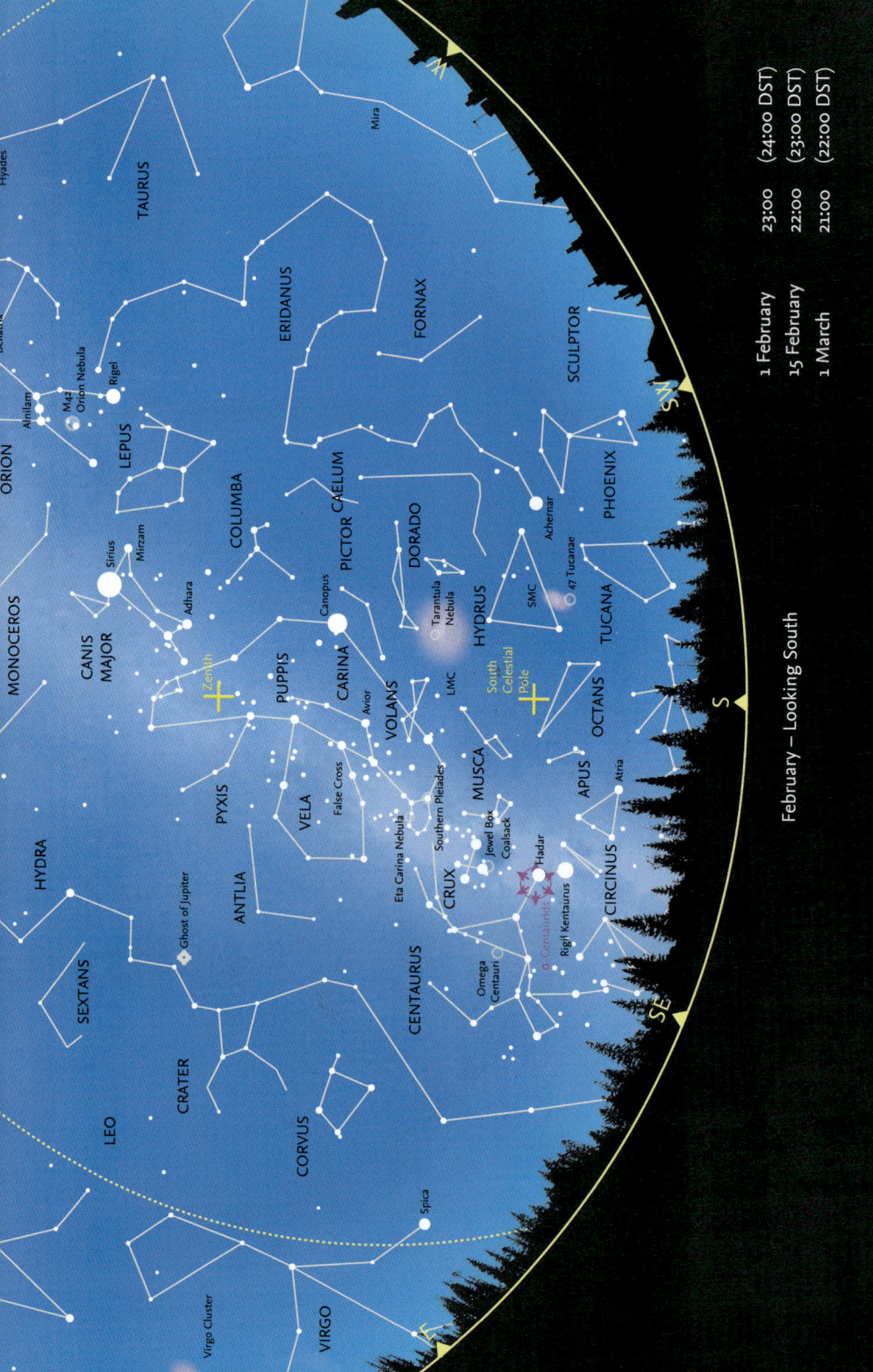

February – Looking South

The whole of *Centaurus* is now clear of the horizon in the southeast, where, below it, the constellation of *Lupus* is beginning to be visible. *Triangulum Australe* is now higher in the sky and easier to see. Both *Crux* and the *False Cross* are also higher and more visible. The *Large Magellanic Cloud* (LMC) and brilliant *Canopus* are now almost due south, with the constellation of *Puppis* at the zenith. Both *Hydrus* and the *Small Magellanic Cloud* (SMC) are lower in the sky, as is *Phoenix*, which is much closer to the horizon in the west. Next to it, the constellation of *Tucana* is also lower although the globular cluster *47 Tucanae* remains readily visible as does *Achernar* (α Eridani). The constellation of *Grus* and bright *Fomalhaut* (α Piscis Austrini) have disappeared below the horizon.

Meteors

The **Centaurid** shower (which actually consists of two separate streams: the α- and β-**Centaurids**), whose radiants lie near α and β Centauri (Rigil Kentaurus and Hadar), respectively – continues in February. The shower reaches a low maximum of around six meteors per hour on 8 February, when the Moon is a bright waning gibbous rising after midnight. Another weak shower, the γ-**Normids**, begins to be active on 25 February, but the meteors are difficult to differentiate from sporadics. It peaks on 14 March, with a similar hourly rate to the Centaurids

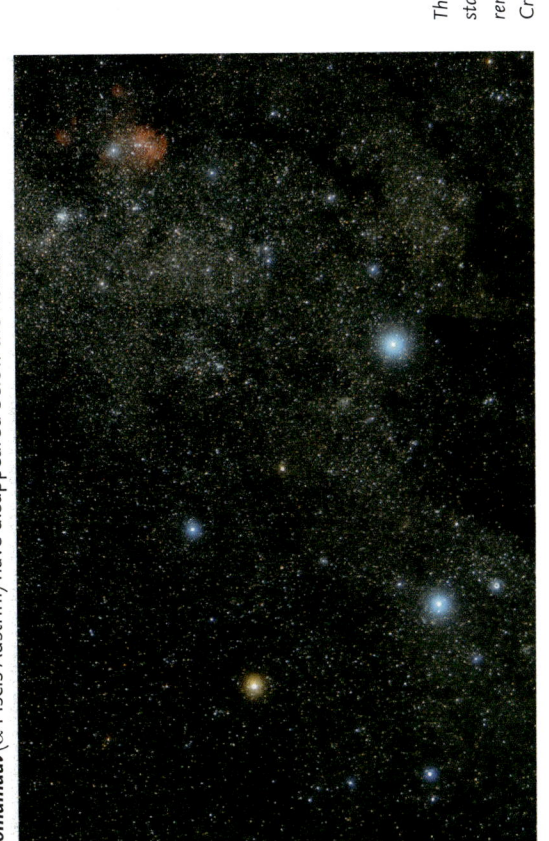

The Southern Cross asterism in the constellation Crux. The brightest star, Acrux (α Crucis) lies to the bottom right of the image. The remaining stars of the Southern Cross are Mimosa bottom left (β Crucis), Gacrux on the left (γ Crucis) and δ Crucis (top).

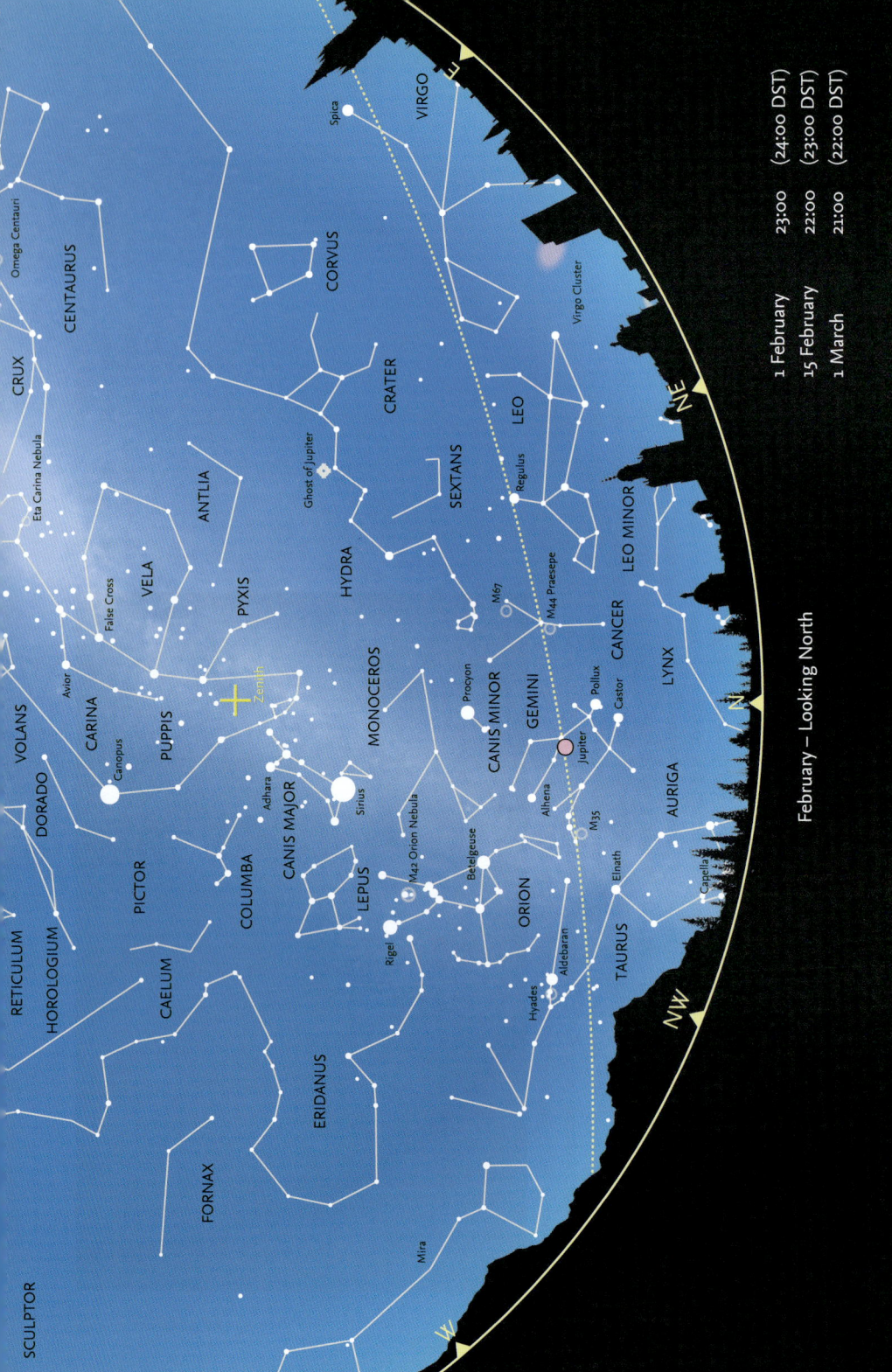

February – Looking North

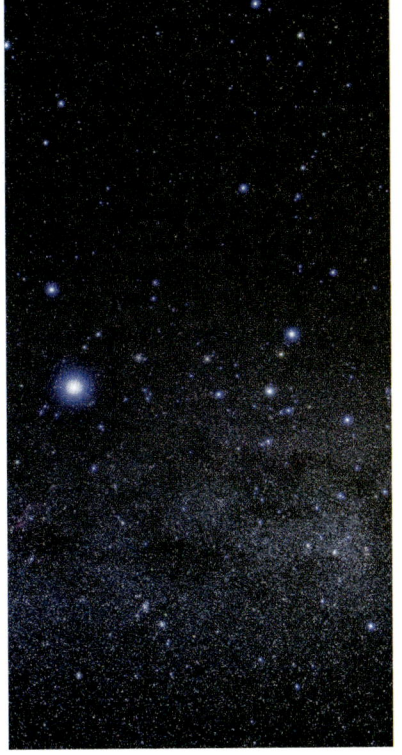

The constellation of **Gemini**, with the pair of stars **Castor** and **Pollux** (α and β Geminorum) is now due north. Pollux is higher in the sky (farther towards the zenith), and above it is **Procyon** (α Canis Minoris), halfway to the zenith. Pollux (β Geminorum) is the brighter of the two (mag. 1.2), it lies in the ecliptic and is occasionally occulted by the Moon. Castor (α Geminorum), the fainter star (mag. 1.9), is remarkable because it is actually a multiple system, consisting of no fewer than six individual stars arranged in three binary (paired) systems.

Still higher is **Sirius**, the brightest star in the sky (at mag. –1.4) and in the constellation of **Canis Major**. The faint constellation of **Cancer**, with its most noticeable feature, the cluster **Praesepe**, lies to the east of Gemini. Above it, and directly east of Procyon, is the distinctive asterism forming the head of **Hydra**, the largest of all 88 constellations, the whole of which is now visible stretching across the sky towards the east. The constellation of **Taurus**, with orange **Aldebaran** (α Tauri) is still clearly seen in the west. By contrast, **Auriga** is much lower towards the horizon and brilliant **Capella** (α Aurigae) is extremely low and visible only early in the night. The faintest stretch of the Milky Way runs from Auriga in the northwest up towards the zenith, passing through Gemini and the constellation of **Monoceros**. In the northeast, the zodiacal constellation of **Leo** and bright **Regulus** (α Leonis) are clearly seen.

The constellation of Canis Major. Brilliant Sirius is prominent.

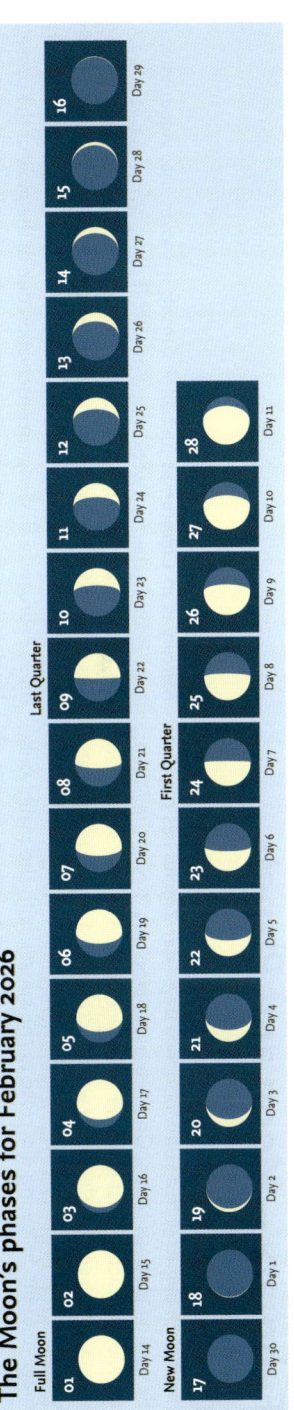

February – Moon and Planets

The Moon

The month begins with a Full Moon, two days later *Regulus* lies 0.4°S of the waning gibbous Moon. On 7 February *Spica* lies 1.8°N of the Moon; the Moon reaches Last Quarter on 9 February. The red supergiant *Antares* can be seen 0.7°N of the waning crescent Moon on 11 February. A New Moon on 17 February is followed by a conjunction with *Mercury* (mag. -0.6) the following day close to the sunset, with the planet only 0.1°N of the thin sliver of waxing crescent Moon. *Saturn* (mag. 1.0) lies 4.6°S of the Moon on 19 February. On 24 February the *Pleiades* lie 1.2°S of the First Quarter Moon. On 27 February *Jupiter* (mag -2.5) will be placed 4.0°S of the waxing gibbous Moon along with *Pollux* 3.0°N of the Moon.

The Planets

Mercury is in *Capricornus* just after sunset; it moves into *Aquarius* and then *Pisces* in the final week of the month. It dims from mag. -1.2 to 1.5. On 18 February Mercury (mag. -0.6) sits only 0.1°N of the thin waxing crescent Moon, the day after it reaches eastern elongation. *Venus* is also present after sunset, at mag. -3.9 all month. It passes 4.7°S of Mercury on 26 February, a challenging observation due to the glow of the sunset. *Mars* starts in Capricornus with Mercury and moves slowly into Aquarius at the end of the month, at mag 1.1. *Jupiter* stays in *Gemini* and can be seen as soon as skies are dark (mag. -2.6 to -2.4). On 27 February it lies 4°S of the waxing gibbous Moon. *Saturn* is in Pisces and observable for a few hours after sunset. On 16 February it appears 1°S of *Neptune*, at mag. 1.0 with Neptune at mag. 7.8. *Uranus* sits in *Taurus* from sunset until midnight (mag. 5-7) and ends its retrograde motion on 4 February. Neptune lies in Pisces with Saturn (mag. 7.8).

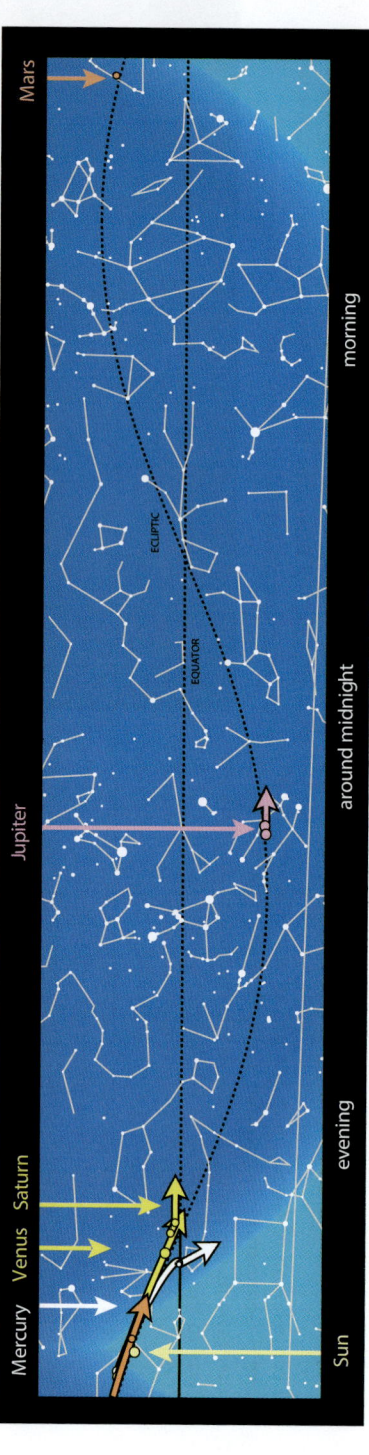

The path of the Sun and the planets along the ecliptic in February.

Calendar for February

01	22:09	Full Moon
03	02:48	Regulus 0.4°S of the Moon
07	08:26	Spica 1.8°N of the Moon
08		α-Centaurids meteor shower maximum
09	12:43	Last Quarter Moon
10	16:52	Moon at apogee = 404,577 km
11	03:19	Antares 0.7°S of the Moon
17	12:01	New Moon
17	12:12	Annular solar eclipse (Antarctica), partial eclipse (south Argentina & Chile, South Africa)
18	23:03	Mercury (mag. -0.6) 0.1°N of the Moon. An occultation will be visible from Mexico, Australia & New Zealand
19	17:39	Mercury at eastern elongation (18.1°E, mag. -0.6)
19	23:54	Saturn (mag. 1.0) 4.6°S of the Moon
24	02:43	Pleiades 1.2°S of the Moon
24	12:28	First Quarter Moon
24	23:18	Moon at perigee = 370,132 km
27	06:26	Jupiter (mag. -2.5) 4.0°S of the Moon
27	21:34	Pollux 3.0°N of the Moon

1–3 February • The Full Moon on 1 February lies in Cancer and passes through Leo as it wanes; it lies next to Regulus on 3 February.

11 February • The waning crescent Moon is next to Antares in Scorpio low in the east at dawn. Sabik is lower towards the north.

19 February • Saturn sits next to the waxing crescent Moon and Mercury (at greatest elongation) at dusk, low in the west.

27 February • The waxing gibbous Moon is next to Pollux (in Gemini) and Jupiter. Procyon and Betelgeuse are also visible.

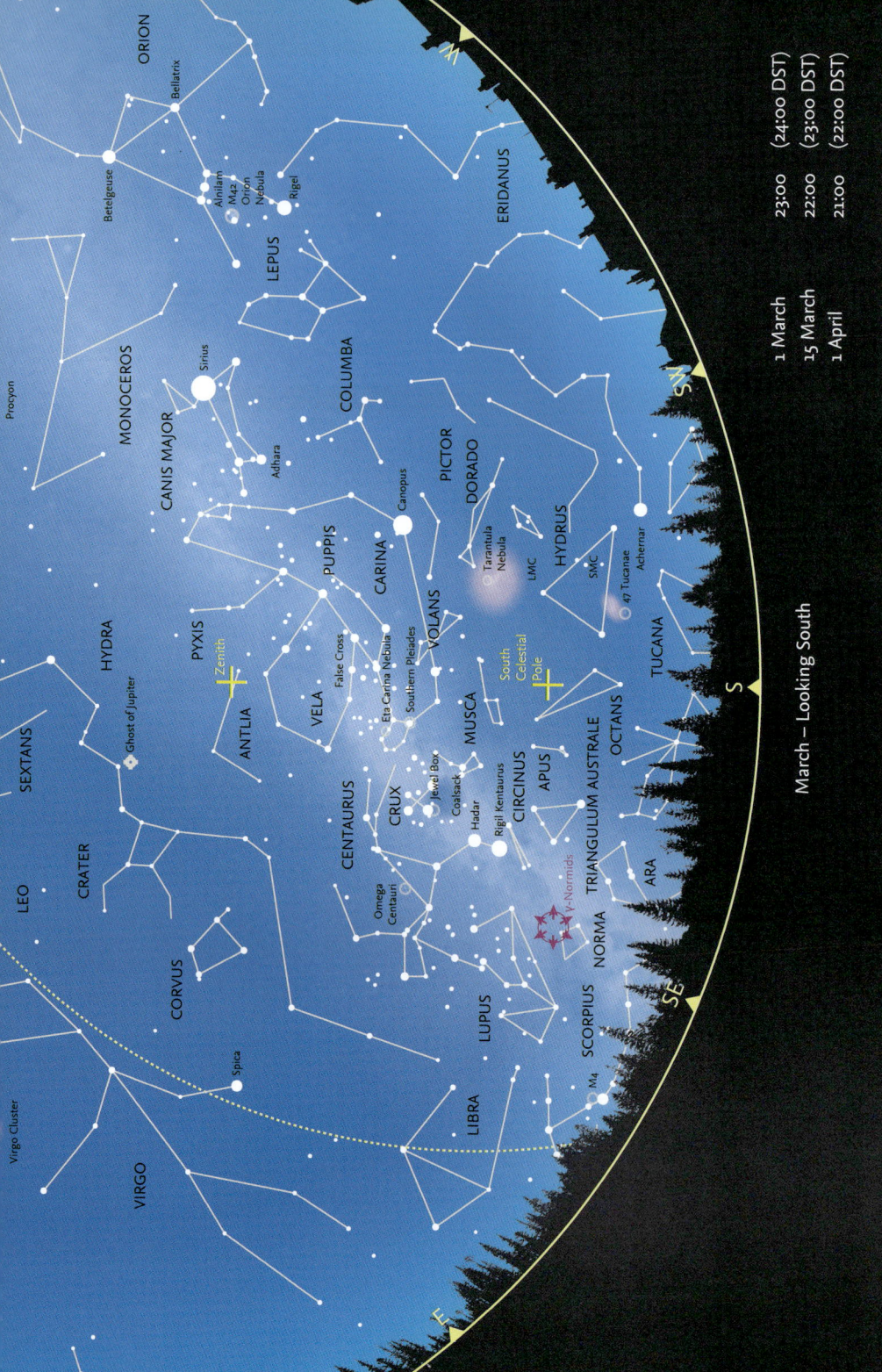

March – Looking South

The Sun crosses the celestial equator on 20 March, at the autumnal equinox, when day and night are of almost equal length, and the southern season of autumn is considered to begin. The hours of daylight change most rapidly around the equinoxes in March and September.

Scorpius is beginning to become visible in the eastern sky, including the Cat's Eyes, the pair of stars **Shaula** and **Lesath** (λ and υ Scorpii, respectively) at the end of the 'sting'. Above Scorpius, the whole of the constellation of **Lupus** is now easy to see. The magnificent globular cluster of **Omega Centauri** is now readily visible, northeast of **Crux**. The **Coalsack Nebula** (a dark nebula of interstellar dust obscuring starlight) and the denser region of the Milky Way in **Carina**, together with the **Eta Carinae Nebula** (containing massive stars considerably hotter than the Sun) and the **Southern Pleiades** are well placed for observation. The **False Cross** on the **Carina/Vela** border is now high in the sky, between the South Celestial Pole and the zenith. Brilliant **Canopus** (α Carinae) is only slightly lower towards the west, above the **Large Magellanic Cloud** (LMC) and the striking **Tarantula Nebula**, a star-forming region visible to the naked eye. **Achernar** (α Eridani), the **Small Magellanic Cloud** (SMC) and **47 Tucanae** are considerably lower, but still clear of the horizon. **Peacock** (α Pavonis), a binary star system in the constellation **Pavo**, skims the horizon, just east of south. **Orion** is now visibly getting lower in the west and is being followed by **Sirius** and **Canis Major**.

Meteors

The meteor shower in March is the γ-**Normids**, which have a low rate, and are thus difficult to differentiate from sporadics. However, they exhibit a very sharp peak a day or so on either side of maximum on 14 March, when the Moon is waning crescent. The faint constellation of **Norma** rises early in the night, but most meteors are likely to be seen (away from the radiant) in the hours after midnight, before the Moon rises.

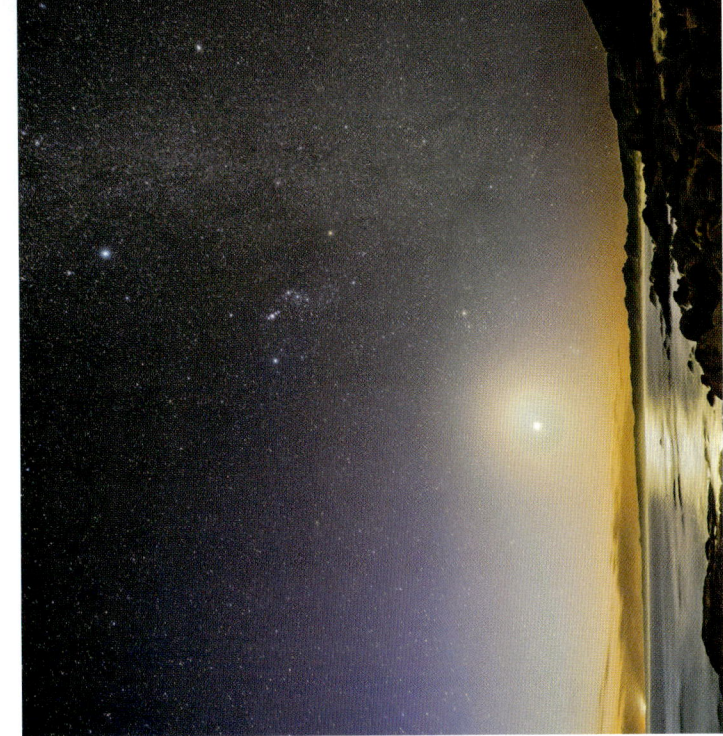

The Milky Way, taken from New Zealand.

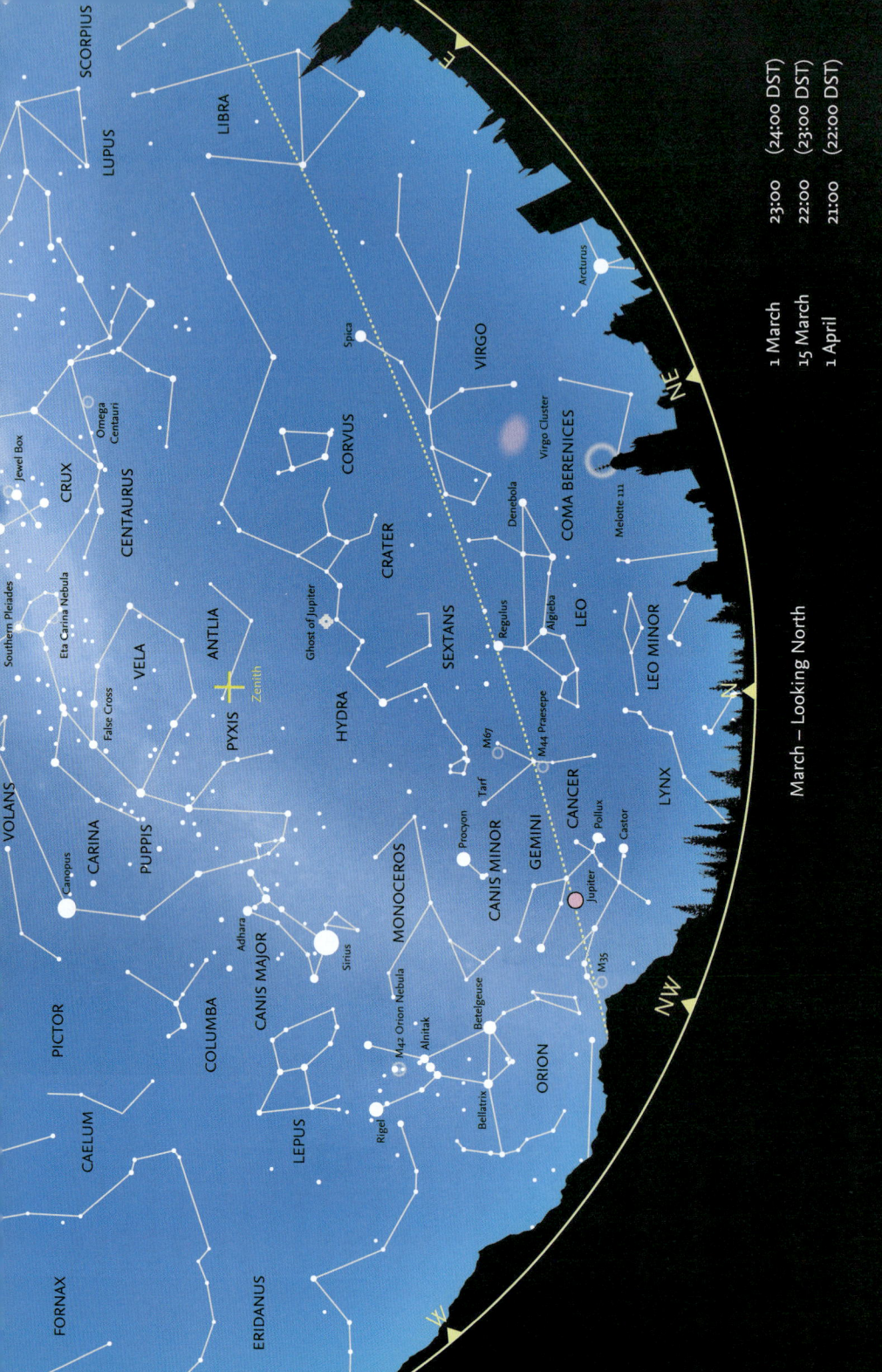

March – Looking North

Almost due north is the constellation of *Leo*, with the 'upside-down backwards question mark' (or the Sickle) of bright stars forming the head of the mythological lion. *Regulus* (α Leonis) – the 'dot' of the question mark or the 'handle' of the sickle and the brightest star in Leo – lies very close to the ecliptic and is one of the few first-magnitude stars that may be occulted by the Moon. *Denebola* (β Leonis), a variable star, marks the tail of the lion. The Moon often passes between Regulus and *Algieba* (γ Leonis). To the west lies the faint zodiacal constellation of *Cancer* (the Crab), its brightest star is β Cancri (also called *Tarf*), an orange giant around 50 times the size of the Sun. Near the centre of the constellation lies an open cluster, M44 or *Praesepe* (the Manger but also known as the Beehive). On a clear night this cluster of around 1,000 stars, known since antiquity, is just a hazy spot to the naked eye, but appears in binoculars as a group of dozens of individual stars. The constellations of *Gemini* and *Orion* are now getting low in the west, but above them, both *Procyon* in *Canis Minor* and the constellation of *Canis Major* remain clear to see. The whole of *Hydra* (the largest constellation) now sprawls right across the northern and eastern skies, and the three small constellations of *Sextans*, *Crater* and *Corvus* are readily visible. In the east, *Arcturus* (α Boötis) – the brightest star in the northern hemisphere of the sky – is becoming visible, and climbs higher during the night. The small constellation of *Antlia* is at the zenith.

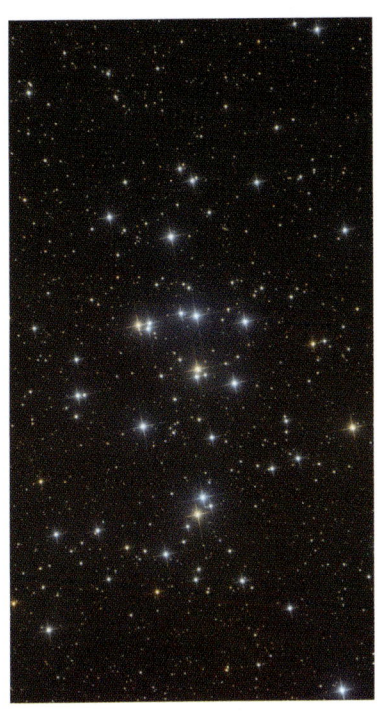

M44 the Beehive Cluster, also known as Praesepe (manger), an open cluster in Cancer.

The Moon's phases for March 2026

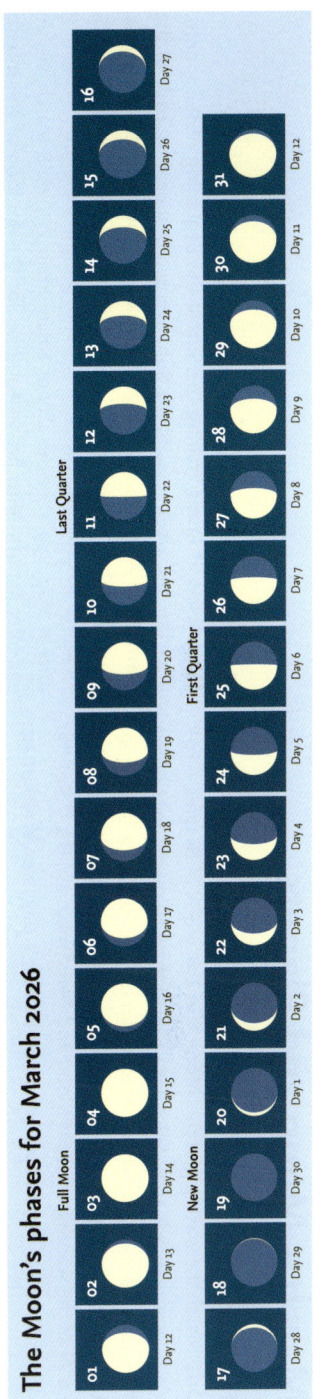

March – Moon and Planets

The Moon

On 2 March the waxing gibbous Moon is in *Leo*, 0.4°N of *Regulus*. A Full Moon occurs on 3 March. *Spica* lies 1.8°N of the waning gibbous Moon three days later. On 10 March *Antares* can be seen 0.7°N of the Moon. A day later the Moon is Last Quarter. On 17 March *Mercury* (mag. 1.7) is 2.0°N of the waning crescent Moon and *Mars* (mag. 1.2) is 1.5°S of the Moon. The New Moon appears on 19 March; a day later *Venus* (mag. -3.9) lies 4.6°S of the very thin crescent Moon. On 23 March, two days before it reaches First Quarter, the Moon sits 1.1°N of the *Pleiades*. On 26 March *Jupiter* (mag. -2.2) passes 3.9°S of the waxing gibbous Moon; the following day *Pollux* appears 3.0°N of the Moon. The bright Moon returns to *Regulus* on 29 March, lying 0.4°N of the brightest star in *Leo*.

The Planets

Mercury lies in *Pisces*, trailing the Sun at the beginning of the month. After the first week it leads the Sun, eventually appearing early in the dawn sky in *Aquarius*, dimming from mag. 2.0 then brightening to 0.4. On 15 March it joins *Mars* (mag. 1.2) lying 3.4°N. *Venus* enters Pisces, dipping into *Cetus* and back as the month progresses and then reaching *Aries* at mag. -3.9. It passes 1°N of *Saturn* (mag. 0.9) on 8 March, visible just after sunset. *Mars* lies in Aquarius for the duration of the month (mag. 1.1 to 1.2). *Jupiter* continues to move slowly through *Gemini* (mag. -2.4 to -2.2). On 11 March it ends retrograde motion and moves eastwards. Saturn is in Pisces, setting shortly after sunset and swinging into the dawn sky after 25 March at mag. 0.9. *Uranus* continues in *Taurus*, dimming from mag. 5.7 to 5.8 and *Neptune* in Pisces approaches the Sun and from 22 March moves into the dawn sky (mag. 7.8 to 7.9).

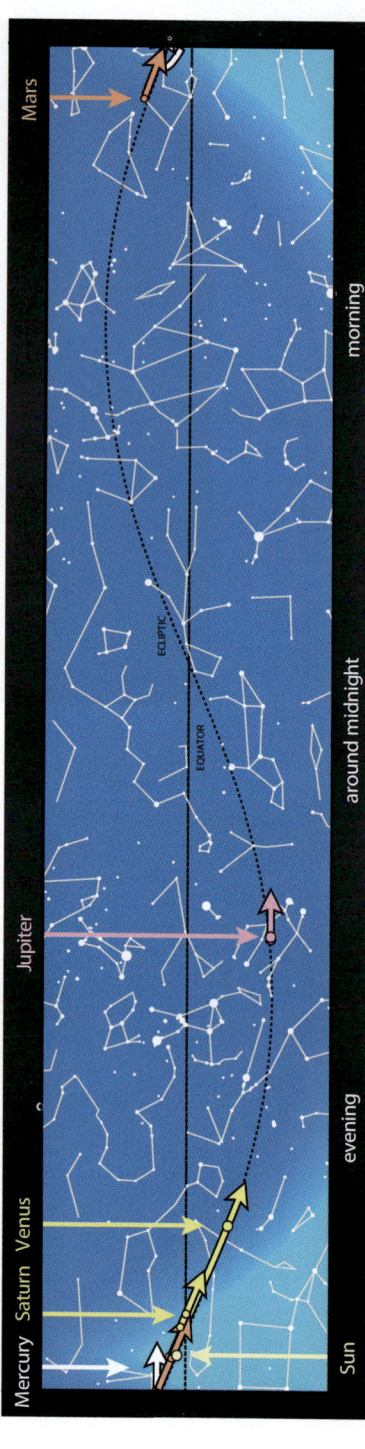

The path of the Sun and the planets along the ecliptic in March.

Calendar for March

02	12:00	Regulus 0.4°S of the Moon
03	11:34	Total lunar eclipse, visible from eastern Asia, Australia, parts of North & South America
03	11:38	Full Moon
06	17:24	Spica 1.8°N of the Moon
10	11:32	Antares 0.7°N of the Moon. An occultation will be visible from Antarctica & New Zealand
10	13:43	Moon at apogee = 404,385 km
11	09:39	Last Quarter Moon
14		γ-Normid meteor shower maximum
15	19:34	Mercury (mag. 2.7) 3.4°N of Mars (mag. 1.2)
17	14:07	Mercury (mag. 1.7) 2.0°N of the Moon
17	21:51	Mars (mag. 1.2) 1.5°S of the Moon
19	01:23	New Moon
20	12:39	Venus (mag. -3.9) 4.6°S of the Moon
20	14:46	Vernal Equinox
22	11:40	Moon at perigee = 366,858 km
23	08:32	Pleiades 1.1°S of the Moon
25	19:18	First Quarter Moon
26	12:13	Jupiter (mag. -2.2) 3.9°S of the Moon
27	03:18	Pollux 3.0°N of the Moon
29	19:00	Regulus 0.4°S of the Moon

2–3 March • The Moon passes close to Regulus on 2 March and becomes Full a day later.

23 March • The Pleiades and the waxing crescent Moon in Taurus.

26 March • The First Quarter Moon is in the vicinity of Jupiter in Gemini. Procyon, Pollux, Castor, Capella, Betelgeuse and Aldebaran are close by.

27 March • Pollux and the waxing gibbous Moon together with Jupiter in Gemini. Capella and Procyon appear either side.

MARCH 51

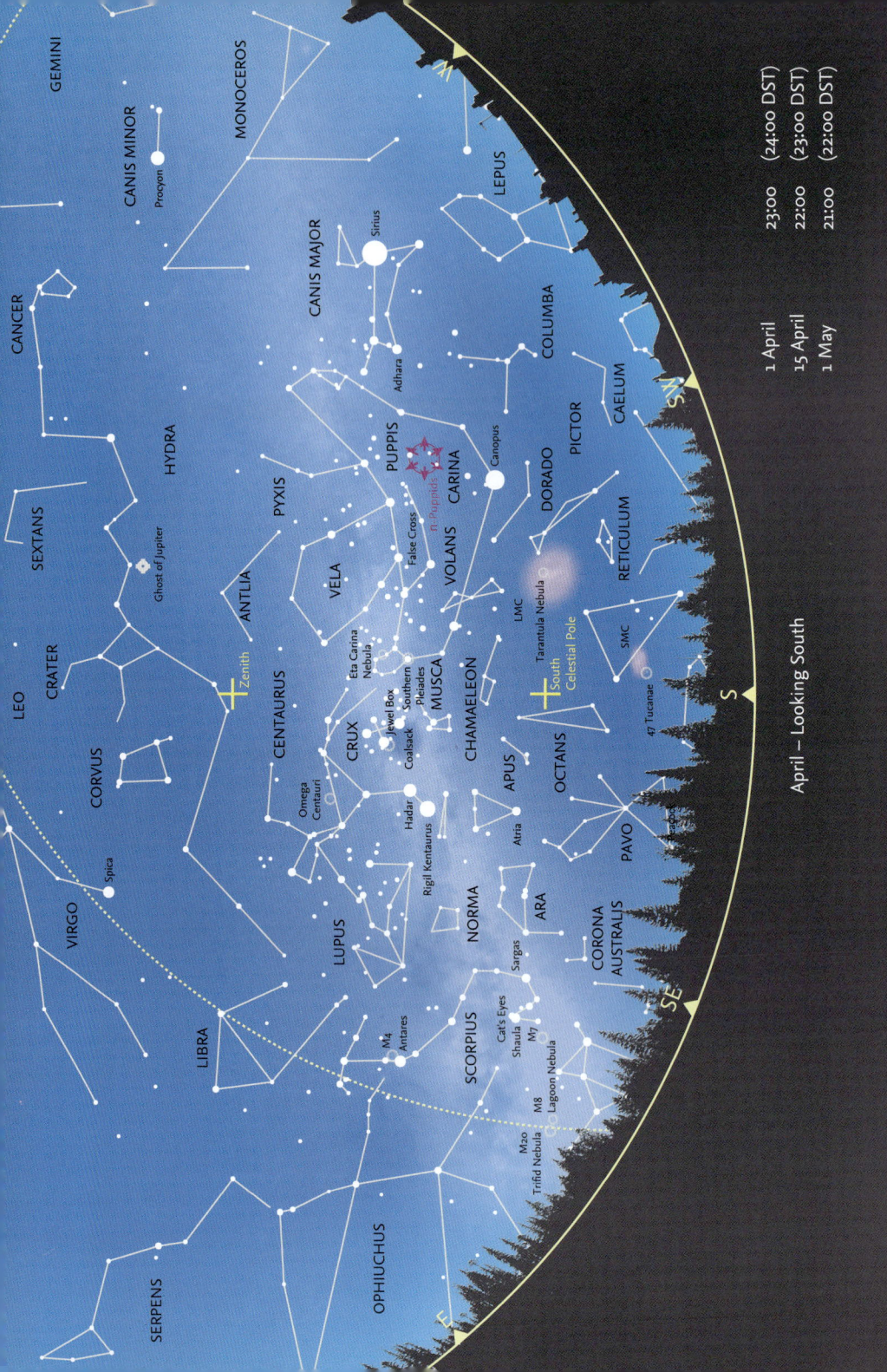

April – Looking South

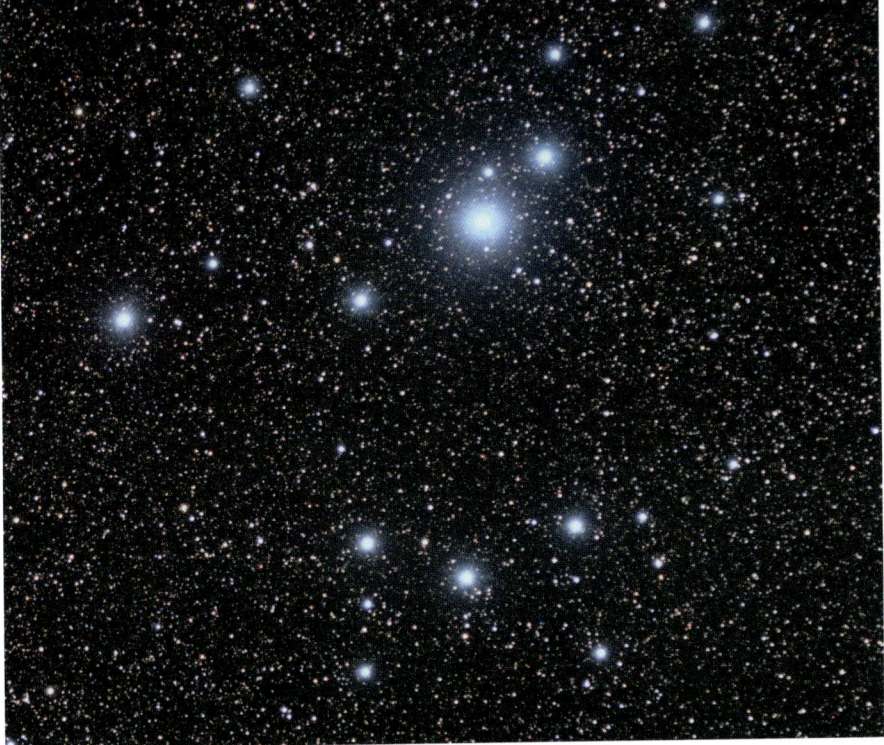

The Southern Pleiades star cluster in the constellation of Carina.

Daylight Saving Time ends where used in Australia and in New Zealand on Sunday 5 April. **Crux** is now high in the south, with the two brightest stars of **Centaurus** (α and β Centauri, **Rigil Kentaurus** and **Hadar**, respectively) to its east. The magnificent globular cluster, **Omega Centauri**, is readily visible high in the sky. West of Crux, both the **Southern Pleiades** and the **Eta Carinae Nebula** are clearly seen, two-thirds of the way towards the zenith. Farther west, the **False Cross** is beginning to decline towards the horizon. **Canopus** (α Carinae) is even lower, and **Orion** has now disappeared below the horizon. **Canis Major** and brilliant **Sirius** are also descending in the west. **Achernar** (α Eridani) is skimming the southern horizon, but **Peacock** (α Pavonis) is now slightly higher and more easily visible. Although the **Large Magellanic Cloud** (LMC) is roughly as high as the South Celestial Pole, the **Small Magellanic Cloud** (SMC) and **47 Tucanae** are rather low (but still visible) in the south. In the east, the whole of **Scorpius** is now well clear of the horizon with the neighbouring zodiacal constellation **Libra** preceding it along the ecliptic. The dense regions of the Milky Way in **Sagittarius** become visible later in the night.

Meteors

Two meteor showers occur in April. The **π-Puppid** shower begins on 15 April; however, there is a First Quarter Moon coinciding with the shower maximum on 24 April so moonlight will interfere with observations. The hourly rate is variable but the meteors tend to be faint. The parent body is the Comet 26P/Grigg-Skjellerup. A second, more prolific, shower, the **η-Aquariids**, begins on 19 April, and continues into May.

APRIL 53

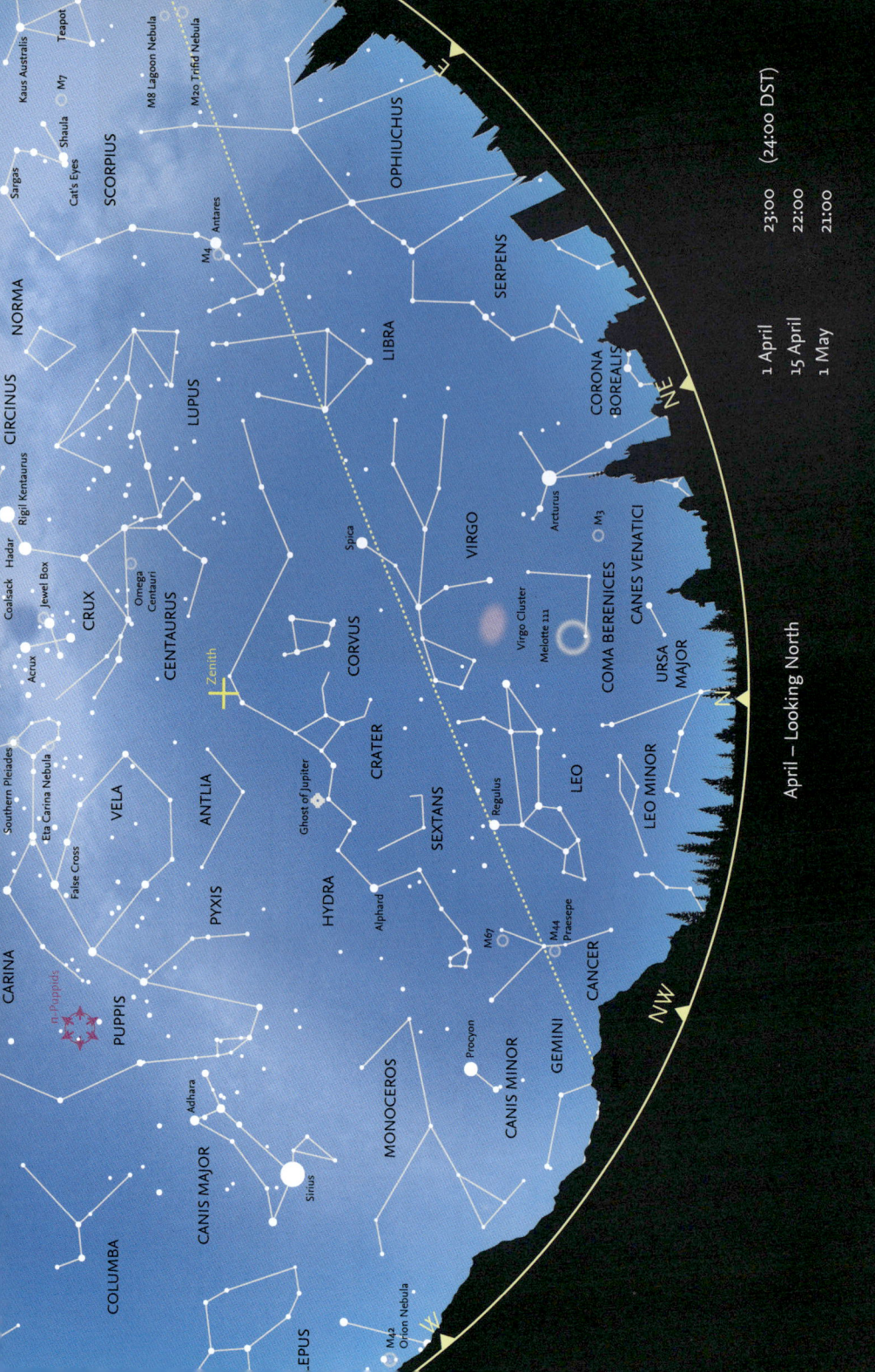

April – Looking North

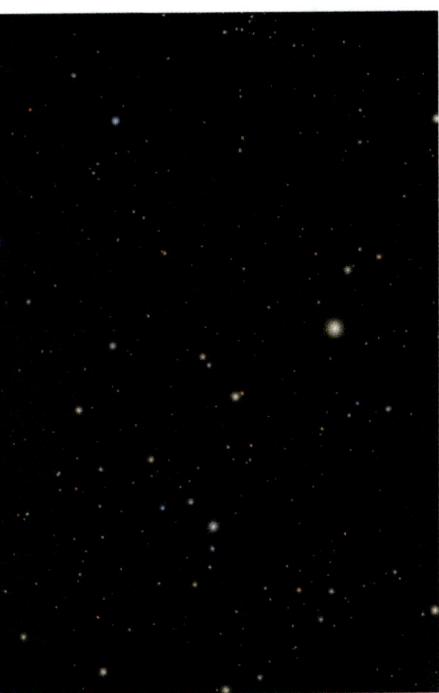

Leo is the most prominent constellation in the northern sky in April and looks vaguely like the creature after which it is named. *Gemini*, with *Castor* and *Pollux*, and *Orion* are low on the horizon in the west, and *Cancer* lies between the two constellations. To the east of Leo, the whole of *Virgo*, with *Spica* (α Virginis) its brightest star, is clearly visible. The constellation of *Libra* is farther east along the ecliptic. Above Leo and Virgo, the complete length of *Hydra* is visible, *Alphard* (α Hydrae) forms a prominent triangle with *Regulus* in Leo and *Procyon* in *Canis Minor*. High in the sky, the small constellations of *Sextans* and *Crater*, together with the rather brighter *Corvus*, lie between Leo, Virgo and Hydra.

Boötes and *Arcturus* are prominent in the northeastern sky, together with the globular cluster M3 in *Canes Venatici*, visible with binoculars or a small telescope. The small circlet of *Corona Borealis* (Northern Crown) is close to the horizon. Between Leo and Boötes lies the constellation of *Coma Berenices*, notable for being the location of the open cluster *Melotte 111* (this is sometimes called the Coma Star Cluster and can be confused with the *Coma Cluster* of galaxies (Abell 1656) east of *Denebola* in *Leo*). There are about 1,000 galaxies in Abell 1656. Only about ten of the brightest galaxies in the Coma Cluster are visible with the largest amateur telescopes.

Boötes with the red giant star Arcturus.

The Moon's phases for April 2026

01 Day 13	02 Day 14	03 Day 15	04 Day 16	05 Day 17	06 Day 18	07 Day 19	08 Day 20
Full Moon							

17 Day 29	18 Day 1	19 Day 2	20 Day 3	21 Day 4	22 Day 5	23 Day 6	24 Day 7
New Moon							First Quarter

09 Day 21	10 Day 22	11 Day 23	12 Day 24	13 Day 25	14 Day 26	15 Day 27	16 Day 28
	Last Quarter						

25 Day 8	26 Day 9	27 Day 10	28 Day 11	29 Day 12	30 Day 13

April – Moon and Planets

The Moon

The Moon is Full on 2 April and as it begins to wane it will pass 1.8°S of *Spica*. On 6 April *Antares* and the waning gibbous Moon will be 0.6° apart, visible a few hours before sunrise. The Last Quarter Moon appears on 10 April. *Mars* (mag. 1.2) and the thin waning crescent Moon are 3.7° apart on 16 April; the following day brings a New Moon. The three-day-old waxing crescent Moon on 19 April lies 4.8°N of *Venus* (mag. -3.9), the *Pleiades* cluster sits 1.0°S of the Moon. On 22 April *Jupiter* (mag. -2.1) can be seen 3.6°S of the Moon; the following day the Moon passes 3.2°S of *Pollux*. On 24 April the Moon is First Quarter. On 26 April the Moon approaches *Regulus*, sitting 0.2°N of the brightest star in *Leo*. The month ends with *Spica* and the waxing gibbous Moon sitting 1.8° apart.

The Planets

Mercury moves from *Aquarius* to *Pisces*, brightening from mag. 0.4 to -0.7. On 3 April Mercury reaches western elongation. Mercury (mag. -0.2) lies 1.7°S of *Mars* (mag. 1.2) and 0.5°S of *Saturn* (mag. 0.9) on 20 April, low on the eastern horizon. *Venus* begins the month in *Aries* and moves into *Taurus* in the latter part of the month. On 8 April Venus (mag. -3.9) sits 4.5°N of (**1**) *Ceres* (mag. 9.0) at dusk and on 24 April it passes within 1°N of *Uranus* (mag. 5.8). Mars moves from Aquarius to Pisces and fluctuates between mag. 1.2 and 1.3. On 19 April it passes 1.2°N of Saturn (mag. 0.9). *Jupiter* stays in *Gemini*, dimming from mag. -2.2 to -2.0. Saturn remains in Pisces and leads the sunrise, pulling away from the Sun as the month progresses (mag. 0.9). Uranus lies in Taurus at mag. 5.8 all month. *Neptune* is in Pisces low in the east before sunrise (mag. 7.8).

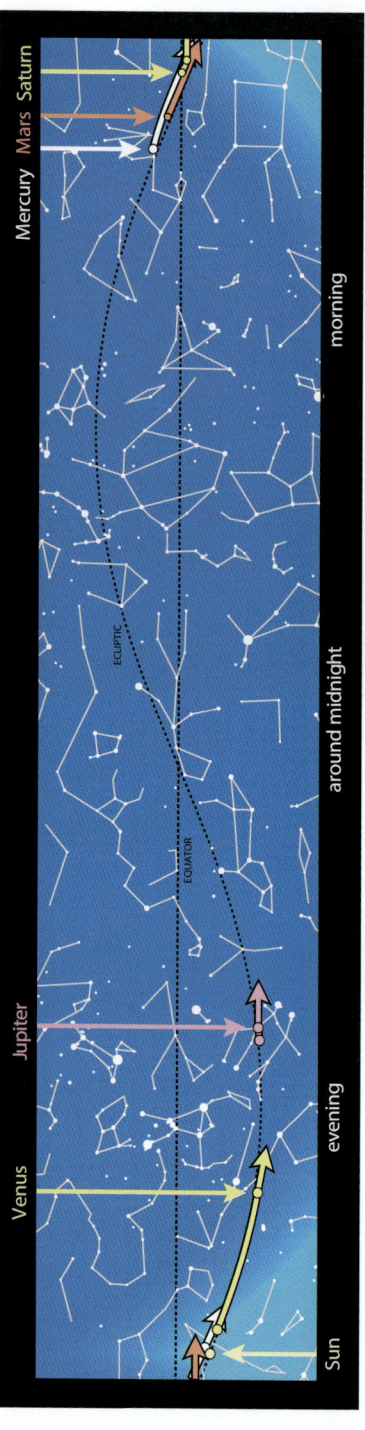

The path of the Sun and the planets along the ecliptic in April.

Calendar for April

01	02:12	Full Moon
03	01:32	Spica 1.8°N of the Moon
03	22:33	Mercury at western elongation (27.8°W, mag. 0.4)
06	19:21	Antares 0.6°N of the Moon. An occultation will be visible from Antarctica, Madagascar, French Southern Territories & Mauritius
07	08:32	Moon at apogee = 404,974 km
10	04:52	Last Quarter Moon
16	00:45	Mars (mag. 1.2) 3.7°S of the Moon
17	11:52	New Moon
19	06:57	Moon at perigee = 361,631 km
19	08:49	Venus (mag. -3.9) 4.8°S of the Moon
19	16:28	Pleiades 1.0°S of the Moon
19	17:36	Mars (mag. 1.2) 1.2°N of Saturn (mag. 0.8)
20	00:00	Mercury (mag. -0.2) 1.7°S of Mars (mag. 1.2)
20	08:03	Mercury (mag. -0.2) 0.5°S of Saturn (mag. 0.8)
22		Lyrid meteor shower maximum
22	22:06	Jupiter (mag. -2.1) 3.6°S of the Moon
23	08:59	Pollux 3.2°N of the Moon
24		π-Puppid meteor shower maximum
24	02:32	First Quarter Moon
24	04:17	Venus (mag. -3.9) 3.4°S of the Pleiades
26	00:37	Regulus 0.2°S of the Moon. An occultation will be visible from Brazil, the eastern Contiguous United States, Colombia & Venezuela
30	08:17	Spica 1.8°N of the Moon

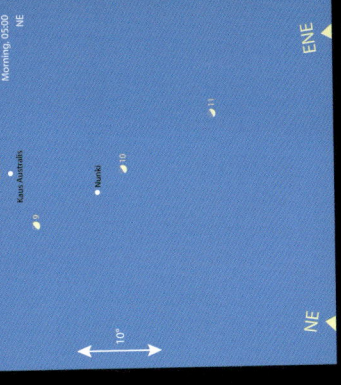

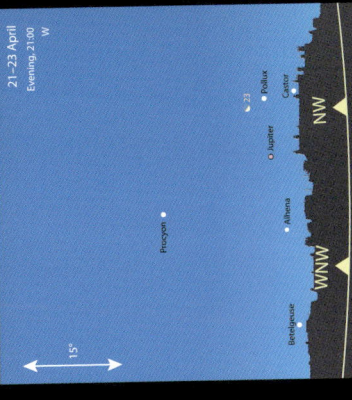

2–3 April • *The Full Moon passes close to Spica, with Arcturus and Denebola in the same region.*

10 April • *Last Quarter Moon in Sagittarius.*

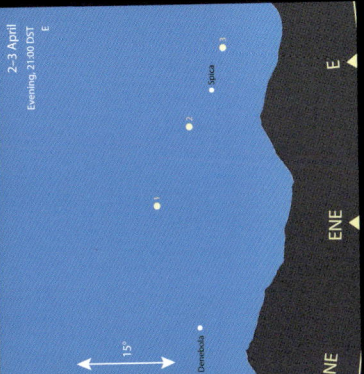

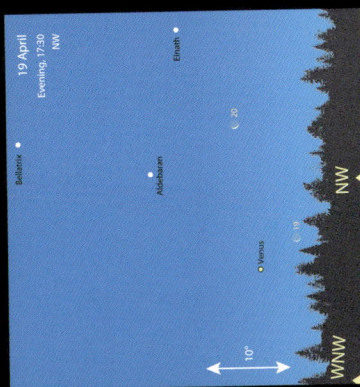

19 April • *Venus and the waxing crescent Moon in Taurus shortly after sunset.*

21–23 April • *The waxing crescent Moon approaching Gemini and Jupiter. Elnath, Capella and Procyon are nearby.*

APRIL 57

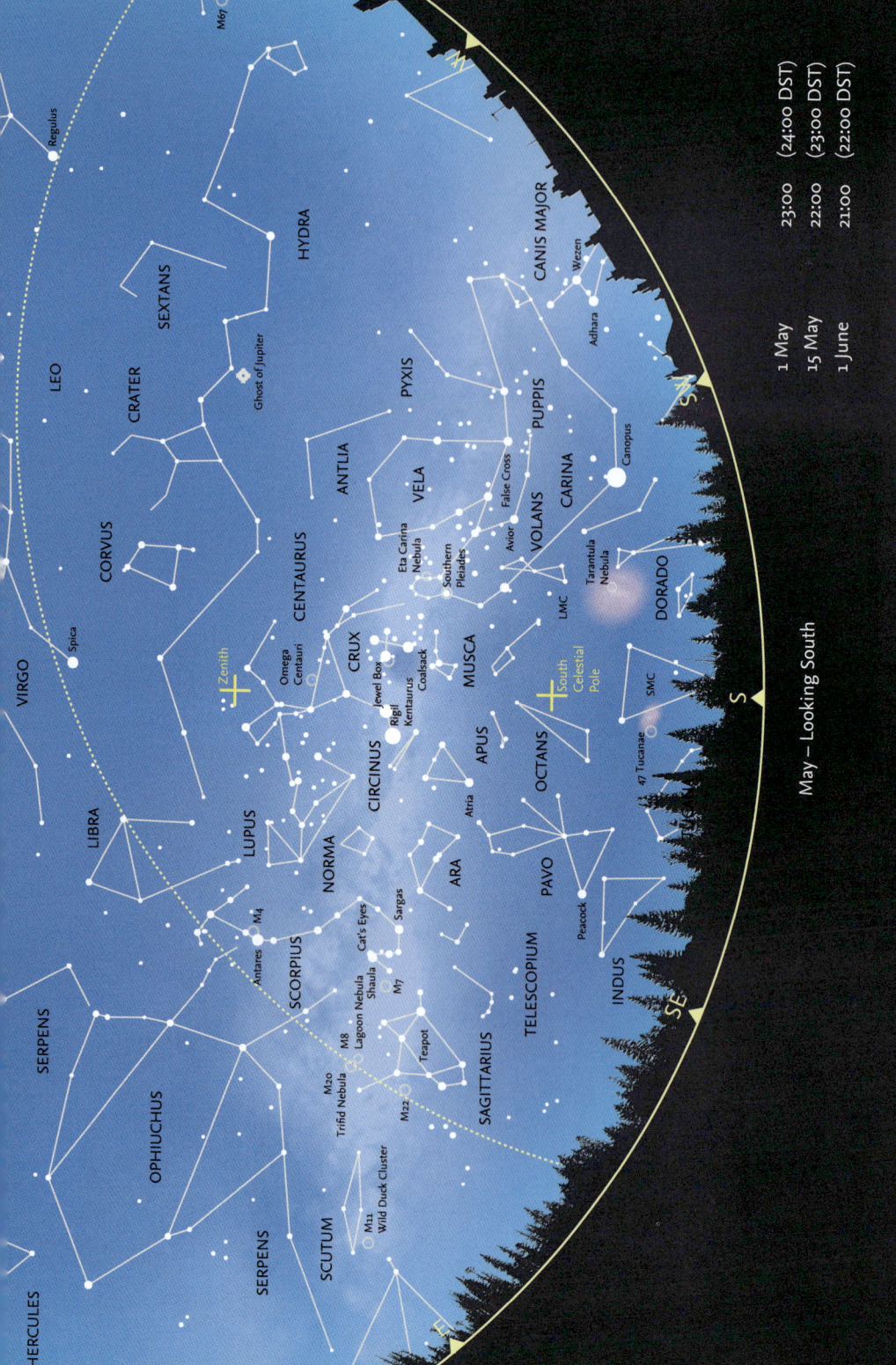

May – Looking South

In the west, **Canis Major** has set below the horizon, and **Canopus** (α Carinae) and **Puppis** are getting rather low. The whole of **Sagittarius** is now clearly seen in the east, with **Corona Australis** beneath it and **Scorpius** higher above. **Centaurus** is fully seen high in the south, with **Crux**, the **Eta Carinae Nebula**, the **Southern Pleiades** and the **False Cross** with **Vela** along the Milky Way to the west. In the south, **Achernar** (α Eridani) is skimming the southern horizon, and the **Small Magellanic Cloud** (SMC) is beginning to rise higher in the sky, unlike the **Large Magellanic Cloud** (LMC), which is now lower. **Pavo** and **Peacock** (α Pavonis) are now much higher and even the faint constellations of **Tucana** and **Indus** are visible between Pavo and the southwestern horizon.

Meteors

The η-**Aquariids** are one of the two meteor showers associated with Comet 1P/Halley, Halley's Comet, last seen in 1986 and not due until 2061. The other shower arising from the cometary debris is the Orionids, in October. The radiant is near the celestial equator, close to the Water Jar in **Aquarius**, well below the horizon. However, meteors may still be seen in the eastern sky even when the radiant is below the horizon. There is a radiant map for the η-Aquariids on page 30.

The peak of the shower on 6 May coincides with a bright waning gibbous Moon. The Moon rises just after midnight; observing conditions will be favourable before then although more meteors are likely to be seen in the early hours of the morning. Maximum hourly rate is about 40 per hour, and a large proportion (about 25 per cent) of the meteors leave persistent trains, which can last for several seconds to minutes.

The constellation Sagittarius (the Archer), with the asterism – the Teapot.

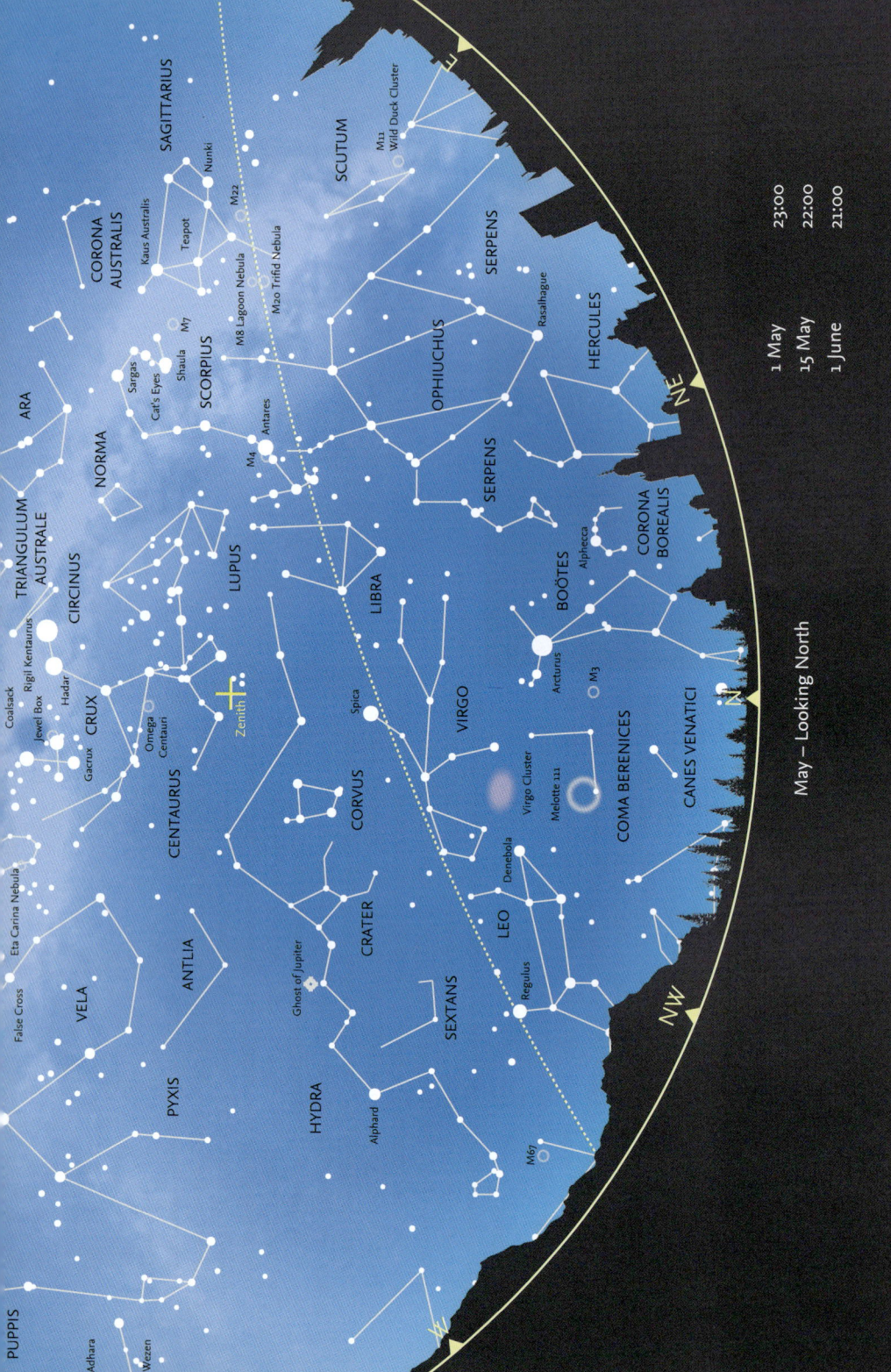

May – Looking North

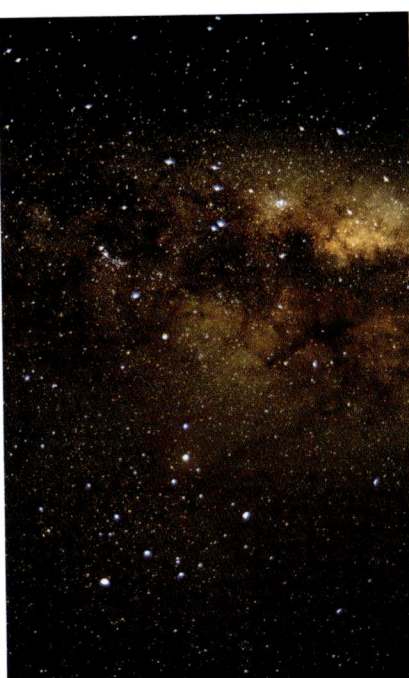

Boötes is almost due north, with brilliant, slightly orange-coloured **Arcturus** extremely prominent. The distinctive circle of **Corona Borealis** is clearly visible to its east. The brightest star (α Coronae Borealis) is known as **Alphecca**. **Hercules** is rising in the east and becomes clearly visible later in the night. To the west of Boötes is **Coma Berenices**, with the cluster **Melotte 111** (see page 55) and, above it, the Virgo Cluster of galaxies. Virgo contains the nearest large cluster of galaxies, which is the centre of the Local Supercluster, of which the Milky Way galaxy forms part. The Virgo Cluster contains some 2,000 galaxies, the brightest of which are visible in amateur telescopes.

The large constellation of **Ophiuchus** (which actually crosses the ecliptic and is thus the 'thirteenth' zodiacal constellation) is climbing into the eastern sky. Before the constellation boundaries were formally adopted by the International Astronomical Union in 1930, the southern region of Ophiuchus was regarded as forming part of the constellation of **Scorpius**, which had been part of the zodiac since antiquity.

Early in the night, the constellation of **Virgo**, with **Spica** (α Virginis), lies due north, with the rather faint constellation of **Libra** to its east.

Scorpius (with red-orange star Antares) taken from Melbourne, Australia, Milky Way visible

Farther along the ecliptic are Scorpius and brilliant, reddish **Antares** (α Scorpii). In the west, the constellation of **Leo** is readily visible now, and both **Regulus** and **Denebola** (α and β Leonis, respectively) are prominent.

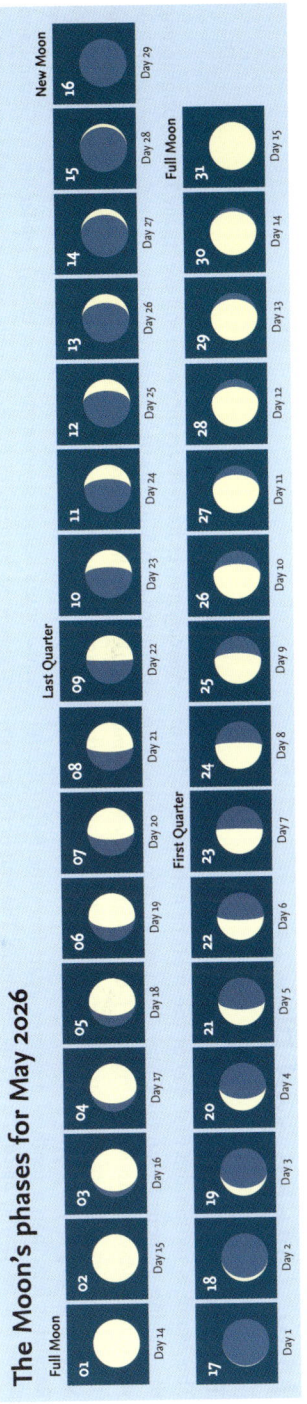

The Moon's phases for May 2026

May – Moon and Planets

The Moon

The Moon is full on the first day of the month. **Antares** is 0.5°N of the waning gibbous Moon on 4 May. The Last Quarter Moon will be visible on 9 May and 16 May brings the New Moon. On 19 May **Venus** (mag. -3.9) lies 2.9°S of the waxing crescent Moon, a day later the Moon swings 3.1°N of **Jupiter** (mag. -1.9) and 3.4°S of **Pollux**. On 23 May **Regulus** is less than a tenth of a degree apart from the First Quarter Moon and on 27 May **Spica** is 1.9°N of the Moon. The end of the month brings a Full Moon with **Antares** 0.4°N.

The Planets

Mercury moves from **Pisces** into **Aries** and eventually **Taurus**, brightening from mag. -0.8 to -2.4 and then dimming to -0.6. In the latter half of the month Mercury appears shortly after sunset. **Venus** starts the month in Taurus and moves into **Gemini**, visible after sunset (mag. -3.9 to -4.0). **Mars** starts in Pisces and passes into Aries, appearing at dawn a few hours before sunrise (mag. 1.2 to 1.3). **Jupiter** lies in Gemini and is above the horizon from sunset until around midnight (mag. -2.0 to -1.9). **Saturn** lies on the border between **Cetus** and Pisces, visible just before sunrise (mag. 0.9). **Uranus** in Taurus approaches the Sun and moves into the dawn sky after 22 May (mag. 5.8). **Neptune** is low in the east before sunrise in Pisces at mag.7.8.

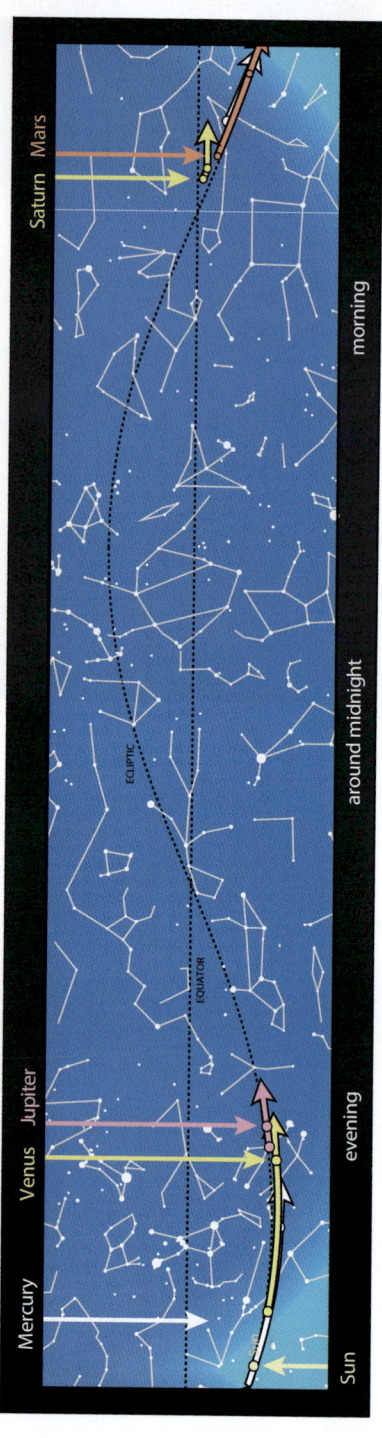

The path of the Sun and the planets along the ecliptic in May.

Calendar for May

01	17:23	Full Moon
04	02:20	Antares 0.5°N of the Moon. An occultation will be visible from Antarctica, Argentina, Chile & Bolivia
04	22:30	Moon at apogee = 405,843 km
06		Eta Aquariid meteor shower maximum
09	21:10	Last Quarter Moon
16	20:01	New Moon
17	13:48	Moon at perigee = 358,074 km
19	01:50	Venus (mag. -3.9) 2.9°S of the Moon
20	12:39	Jupiter (mag. -1.9) 3.1°S of the Moon
20	16:30	Pollux 3.4°N of the Moon
23	06:41	Regulus 0.0°N of the Moon
23	11:11	First Quarter Moon
27	14:09	Spica 1.9°N of the Moon
31	08:32	Antares 0.4°N of the Moon. An occultation will be visible from Argentina, eastern Australia, Chile & New Zealand
31	08:45	Full Moon

6 May • *The waning gibbous Moon in Sagittarius in the eastern sky, the Eta Aquariids meteor shower takes place in the dawn sky.*

19–21 May • *The waxing crescent Moon in Gemini with Jupiter and Venus, Pollux and Castor, Procyon and Capella lie close by.*

23 May • *The First Quarter Moon close to Regulus and Denebola in Leo.*

30 May • *Saturn rising in the east along with Mars just before sunrise. Alpheratz (α And) sits lower down.*

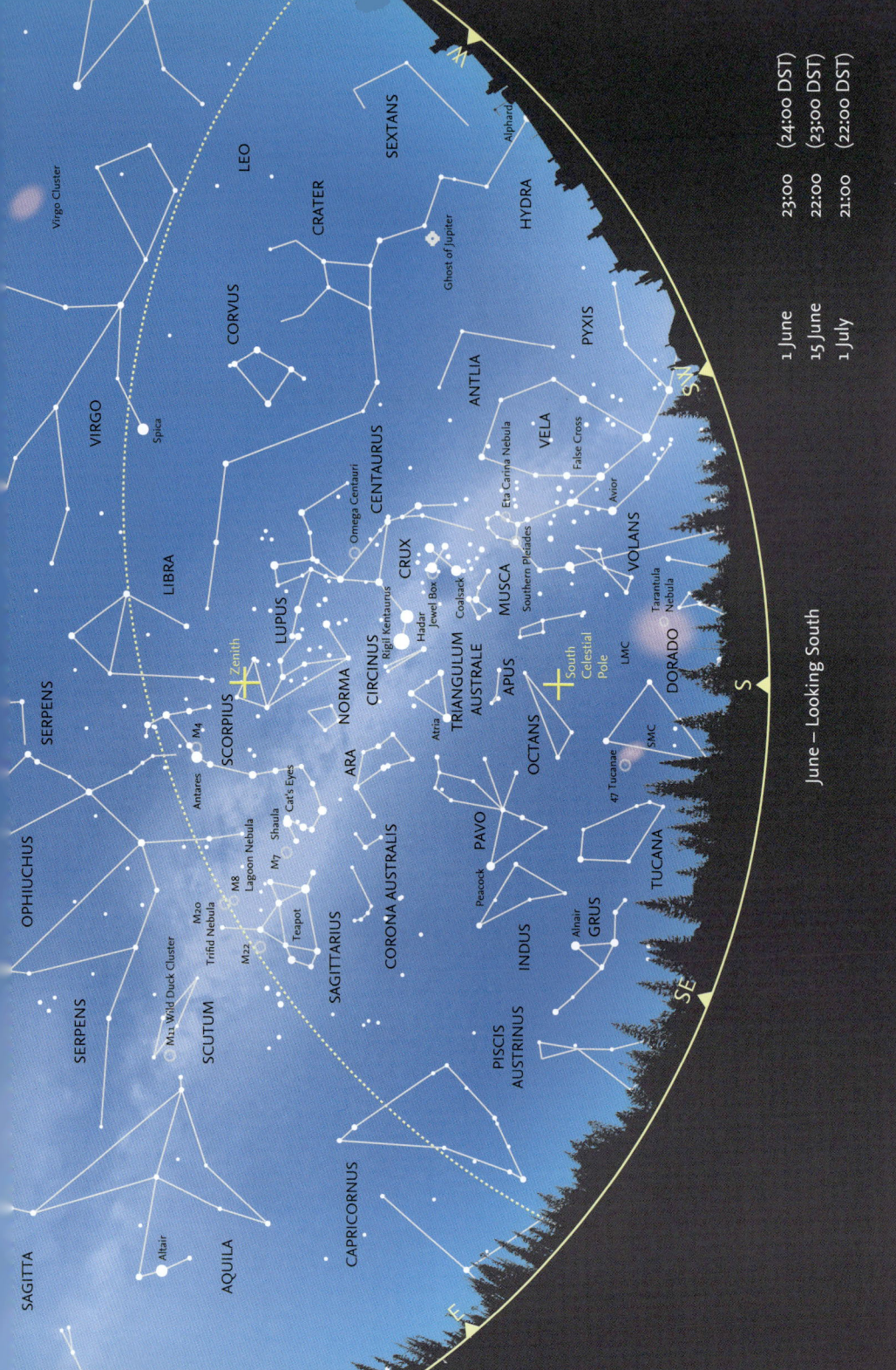

June – Looking South

The winter solstice on 21 June brings the shortest period of daylight. Both *Canopus* (α Carinae) and *Achernar* (α Eridani) are skimming the southern horizon. Although the whole of *Carina* is visible, all of *Eridanus* (except *Achernar*) is hidden below the horizon. The *Large Magellanic Cloud* (LMC) is low, although the *Small Magellanic Cloud* (SMC), the globular cluster, *47 Tucanae* and *Hydrus* are now rather higher. *Alphard* (α Hydrae) is on the horizon, and the constellation of *Sextans* is becoming low. However, the remainder of the long constellation of Hydra is clearly seen as are the two constellations of *Crater* and *Corvus* to its north. The *False Cross* between Carina and *Vela* is beginning to descend in the west, but Vela itself is clearly visible. The whole of both *Crux* and *Centaurus* are clearly seen, as are the magnificent globular cluster, *Omega Centauri*, and the constellation of *Lupus*, closer to the zenith. The constellation of *Triangulum Australe* is on the meridian, roughly halfway between the South Celestial Pole and the zenith. Both *Scorpius*, with brilliant, red *Antares* (α Scorpii) and *Sagittarius* are high overhead, with the faint constellation of *Corona Australis* visible below them. The whole of *Capricornus* is visible, and the constellation of *Grus* has now risen above the horizon, with the faint constellation of *Indus* between it and *Pavo*. To the east of Grus is *Piscis Austrinus*, although brilliant *Fomalhaut* (α Piscis Austrini) is only just clear of the horizon and becomes clearly visible only later in the night and later in the month.

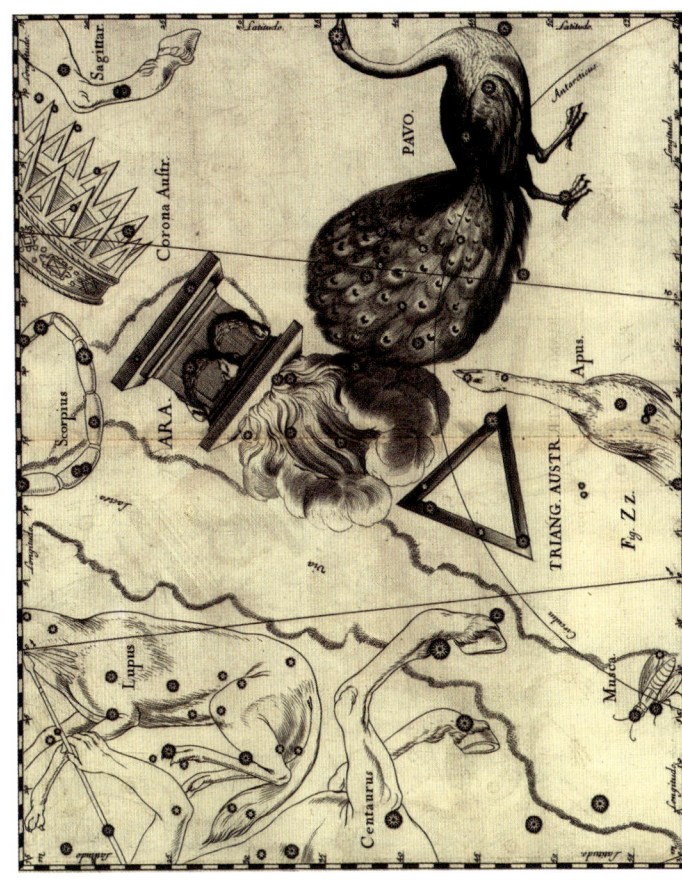

The constellation Pavo, with the Peacock star (α Pavonis).

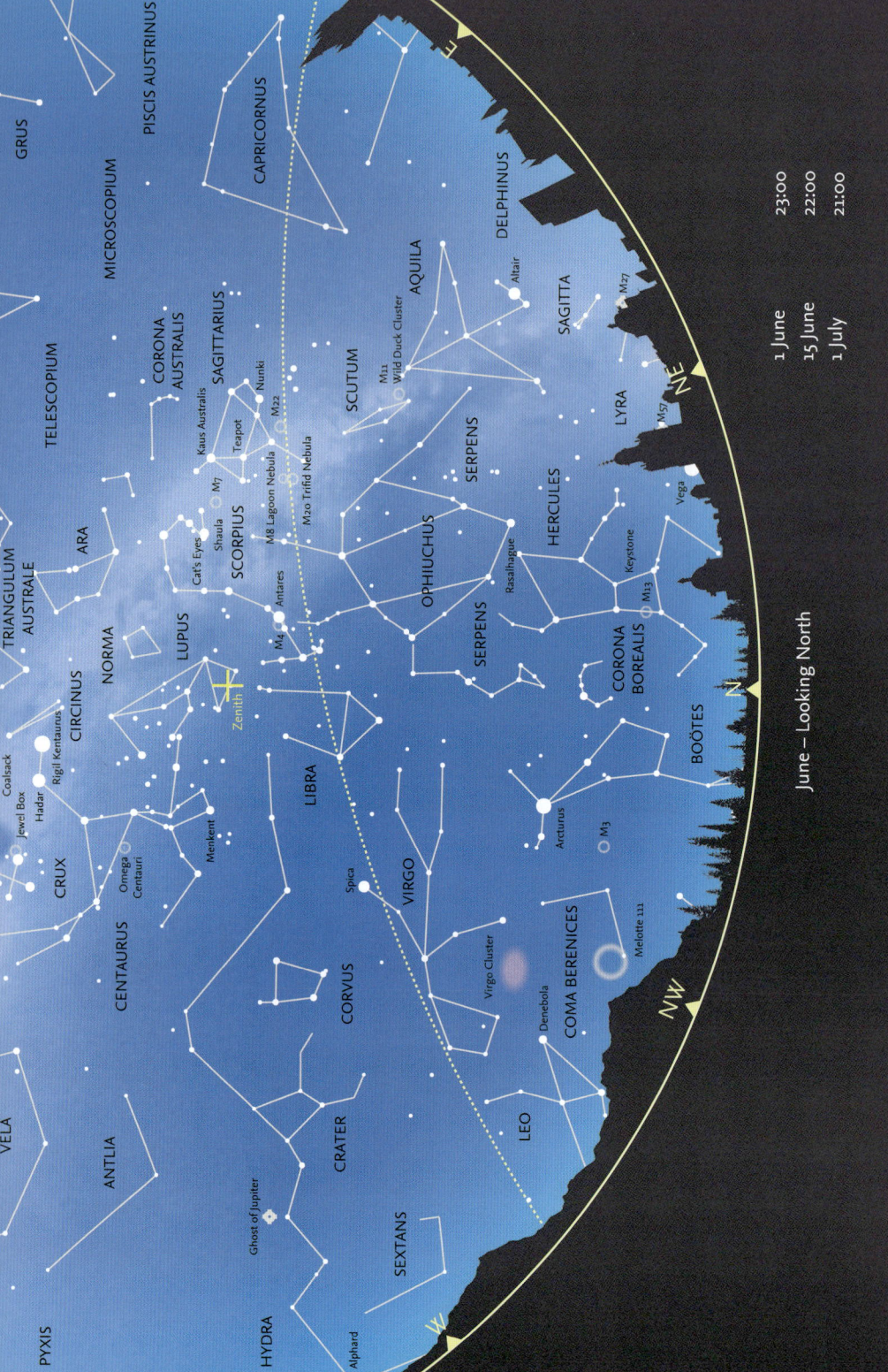

June – Looking North

Vega in **Lyra** is just above the northeastern horizon. Closer to the meridian, the whole of **Hercules** is visible including the Keystone and **M13**, widely regarded as the finest globular cluster in the northern hemisphere. The small constellation of **Corona Borealis** is almost due north. To the west of the meridian is **Boötes** and **Arcturus** (α Boötis). Much of **Leo** is now below the western horizon, but **Denebola** (β Leonis) is still visible. Higher in the sky, the whole of **Virgo** is clearly visible and, still higher, not far from the zenith is the constellation of **Libra**. To its east is **Scorpius** and the red supergiant **Antares** – the name means the 'rival of Mars'. Farther along the ecliptic, both the constellations of **Sagittarius** and **Capricornus** are completely visible. Between Hercules and Sagittarius is the large constellation of **Ophiuchus** (the Serpent Bearer), lying between the two halves of the constellation of **Serpens**: **Serpens Caput** (Head of the Serpent) to the west and **Serpens Cauda** (Tail of the Serpent) to the east. The ecliptic runs across Ophiuchus and, to its east, **Aquila** and **Altair** (α Aquilae), one of the stars of the (northern) Summer Triangle.

Ophiuchus and Serpens. The ecliptic passes through Ophiuchus.

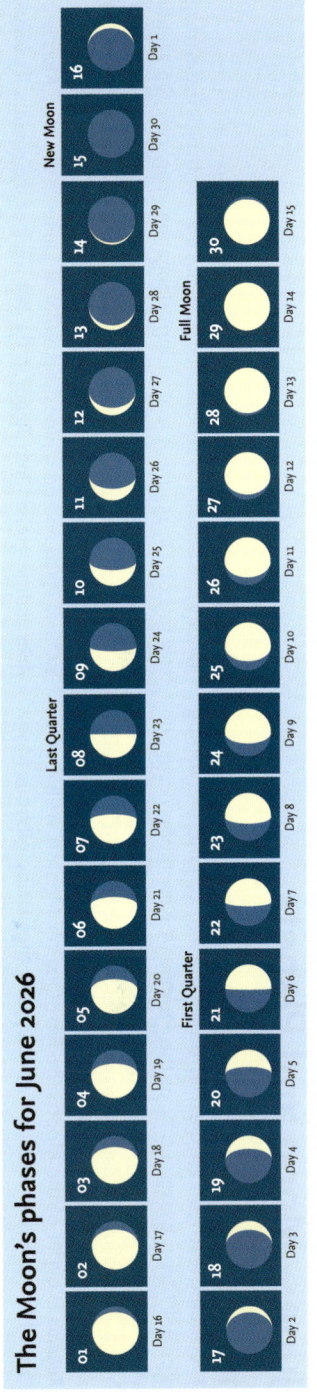

The Moon's phases for June 2026

June – Moon and Planets

The Moon

The Moon reaches Last Quarter on 8 June. On 13 June the waning crescent Moon passes 1.0°N of the *Pleiades*. A New Moon occurs on 15 June; a day later *Mercury* (mag. 0.5) lies 2.6°S of a thin sliver of the Moon. The following day the waxing crescent Moon is 3.6°S of *Pollux*, 2.5°N of *Jupiter* (mag. -1.8) and 0.3°N of *Venus* (mag. -4.0). On 19 June *Regulus* is 0.3°N of the Moon. The First Quarter Moon is visible on 21 June at the summer solstice. Two days later the waxing gibbous Moon passes 2.2°S of *Spica*. On 27 June *Antares* lies 0.5°N of the Moon and on 29 June the Moon is Full.

The Planets

Mercury is in *Gemini* and appears after sunset as the month progresses (mag. -0.5 to 2.0). On 15 June it is at eastern elongation (mag. 0.5). On 25 June Mercury (mag. 1.3) passes 3.8°S of *Jupiter* (mag. -1.8). *Venus* starts the month in *Gemini*, 12 days later it moves into *Cancer* and by the end of the month it is in *Leo* and visible in the evening (mag. -4.0 to -4.1). On 7 June Venus is 4.6°S of *Pollux* and on 9 June it sits 1.6°N of *Jupiter* (mag. -1.9). *Mars* is in *Aries* and progresses into *Taurus* in the last week of the month. It's visible up to a few hours before sunrise (mag. 1.3 to 1.4). On 28 June Mars is 4.3°S of the *Pleiades* (mag. 1.4). Jupiter moves through Gemini and into Cancer, visible from sunset until around 22:00 (mag. -1.9 to -1.8). *Saturn* sits in *Pisces*, appearing a few hours after midnight (mag. 0.9 to 0.8). *Uranus* is in *Taurus* and only observable around an hour before sunrise (mag. 5.8). *Neptune* lies in Pisces and above the horizon several hours after midnight (mag. 7.8).

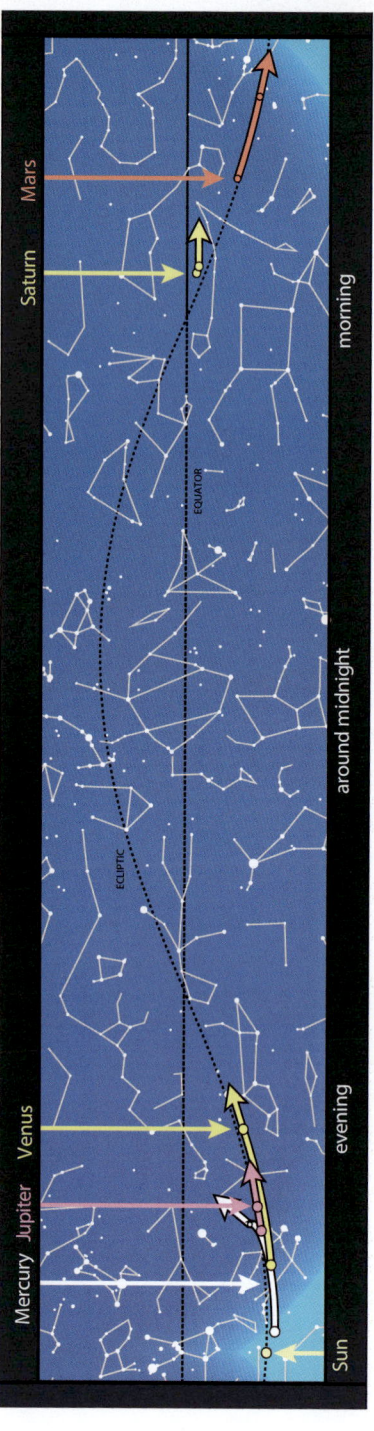

The path of the Sun and the planets along the ecliptic in June.

Calendar for June

01	04:32	Moon at apogee = 406,369 km
07	16:17	Venus (mag. -4.0) 4.6°S of Pollux
08	10:00	Last Quarter Moon
09	20:03	Venus (mag. -4.0) 1.6°N of Jupiter (mag. -1.9)
13	13:15	Pleiades 1.0°S of the Moon
14	23:18	Moon at perigee = 357,196 km
15	02:54	New Moon
15	20:00	Mercury at eastern elongation (24.5°E, mag. 0.5)
16	19:32	Mercury (mag. 0.5) 2.6°S of the Moon
17	02:08	Pollux 3.6°N of the Moon
17	06:54	Jupiter (mag. -1.8) 2.5°S of the Moon
17	20:21	Venus (mag. -4.0) 0.3°S of the Moon. An occultation will be visible from Canada, Brazil & Venezuela
19	14:31	Regulus 0.3°N of the Moon. An occultation will be visible from South Africa, Mozambique, Madagascar & Zimbabwe
21	08:25	Summer Solstice
21	21:55	First Quarter Moon
23	20:11	Spica 2.2°N of the Moon
25	12:00	Mercury (mag. 1.3) 3.8°S of Jupiter (mag. -1.8)
27	14:32	Antares 0.5°N of the Moon. An occultation will be visible from Antarctica, southeastern Australia, New Zealand & Tasmania
28	07:11	Moon at apogee = 406,267 km
28	18:32	Mars (mag. 1.3) 4.3°S of the Pleiades
29	23:57	Full Moon

9 June • *Venus and Jupiter appear close together in Gemini, with Mercury nearby. Pollux and Castor shine brightly.*

19–23 June • *The Moon progresses eastward from Regulus to Spica, reaching First Quarter on 21 June.*

27 June • *The waxing gibbous Moon lies close to Antares, the rival of Mars. Sabik (η Oph) lies near by.*

29 June • *Mars, Uranus and the Pleiades all in Taurus rising in the east just before sunrise.*

JUNE

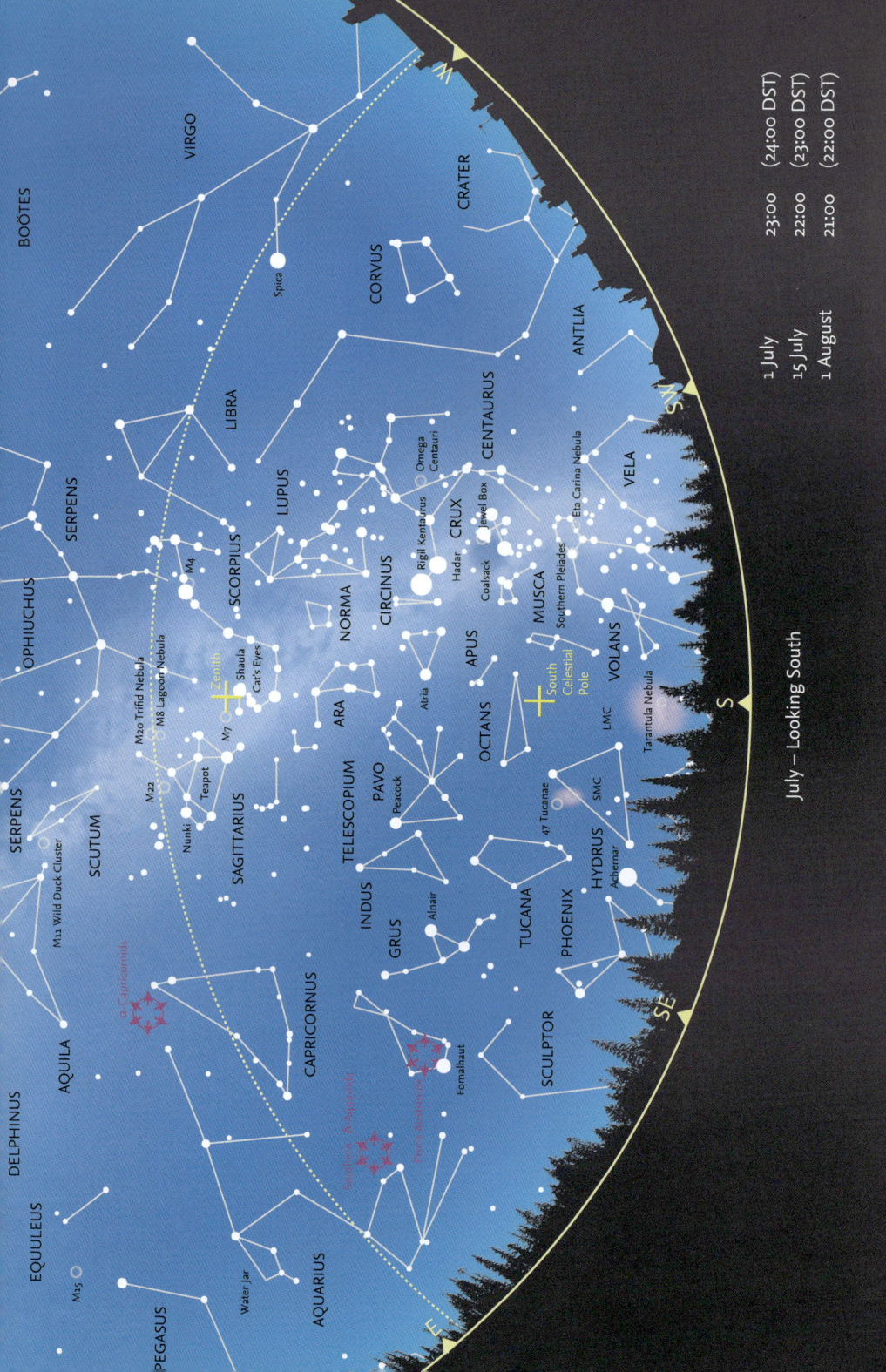

July – Looking South

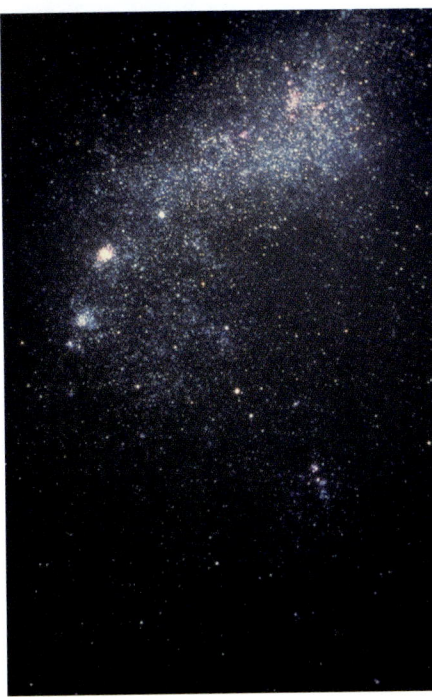

The Small Magellanic Cloud, a dwarf galaxy of several hundred million stars, in the constellation Tucana.

Although the *Large Magellanic Cloud* (LMC) is almost due south, it is very low. Slightly farther west, the *False Cross* is nearing the horizon. Some of *Hydra* remains visible, but the constellation of *Crater* is becoming very low. *Corvus* is still easy to see. The zodiacal constellation of *Virgo* will soon be disappearing in the west. *Crux*, *Centaurus* and *Lupus* are still clearly seen, high in the sky, as is *Scorpius*, the tail of which is near the zenith. *Achernar* (α Eridani), *Hydrus* and the *Small Magellanic Cloud* (SMC) are now higher above the horizon and easier to observe. The whole of the constellation of *Phoenix* is also now clear of the horizon. Above it are *Tucana* and, halfway to the zenith, the constellation of *Pavo*. *Grus* and *Piscis Austrinus* (with brilliant *Fomalhaut*) are now fully visible. Higher in the sky are the zodiacal constellations of *Capricornus* and *Aquarius*, and even the westernmost portion of *Pisces* is rising above the horizon.

Meteors

July brings three meteor showers; there are two minor radiants active in the constellations of *Capricornus* and *Aquarius*. The first shower, the α-*Capricornids*, active from 3 July–15 August (peaking 30 July), has a maximum rate of about five per hour; however, it does often produce very bright fireballs. The parent body is Comet 169/NEAT. The most prominent shower is probably that of the *Southern* δ-*Aquariids*, which are active from around 12 July–23 August, also with a peak on 30 July, although even then the hourly rate is unlikely to reach 25 meteors per hour. In this case, the parent body is possibly Comet 96P/Machholz. This year, both shower maxima occur when the Moon is a bright waning gibbous, meaning moonlight will interfere with observations. The *Piscis Austrinids* begin on 15 July and continue until 10 August. Maximum is on 28 July, but the rate is only about five per hour. The *Perseids* begin on 17 July and peak on 13 August.

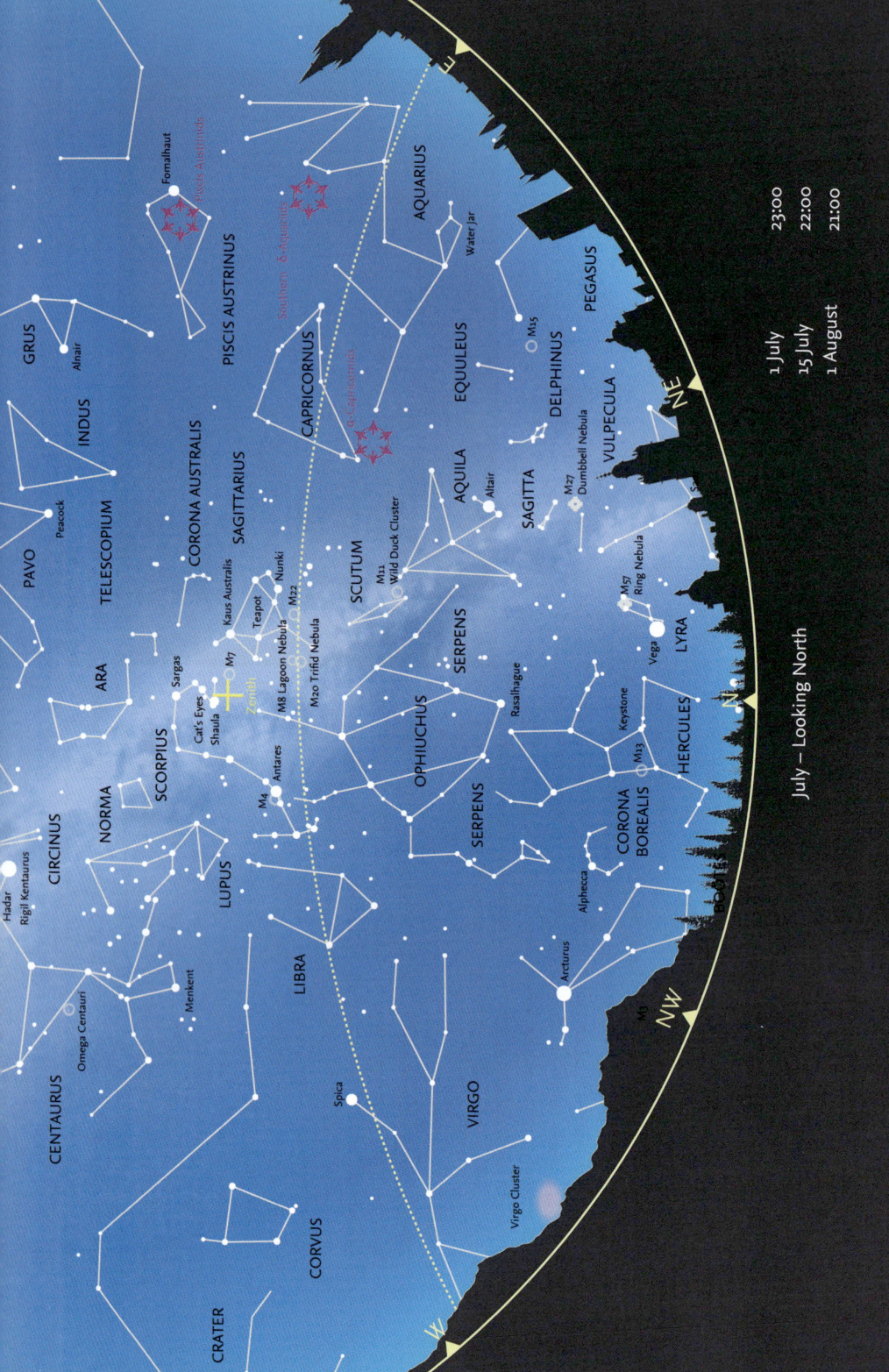

July – Looking North

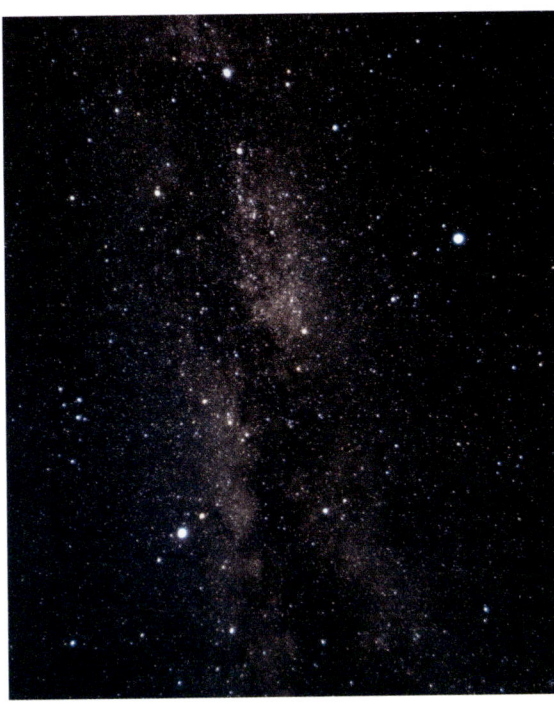

The constellations of *Hercules* and *Lyra* are on either side of the meridian and both are clearly visible well above the horizon. To the east, **Deneb** (α Cygni) is skimming the horizon, but nearly all the rest of the constellation is easily seen, as is **Aquila** with **Altair** (α Aquilae). Between **Cygnus** and Aquila lies the small constellation of *Sagitta*, with **M27** (the Dumbbell Nebula, a planetary nebula). Even farther east, the whole extent of both the zodiacal constellations of **Aquarius** and **Capricornus** is visible. Above Aquila, farther along the Great Dark Rift – dust clouds that hide the more distant stars – is the small constellation of *Scutum*, with **M11**, the Wild Duck Cluster. To its east is the tiny, but distinctive constellation of *Delphinus*. Still farther along the Great Rift lies the centre of the Milky Way Galaxy (in *Sagittarius*) and, near it, two emission nebulae: **M8** (the Lagoon Nebula) and **M20** (the Trifid Nebula). In the western sky, the constellation of **Boötes** and **Arcturus** (α Boötis) are beginning to approach the horizon, but *Corona Borealis* is still clearly seen. Even farther west, the whole of **Virgo** is visible, with **Libra** above it. The large constellation of **Ophiuchus** and the two halves of **Serpens** lie between Hercules and the zenith. *Scorpius* is draped around the zenith with Sagittarius and the distinctive asterism of the Teapot to its east.

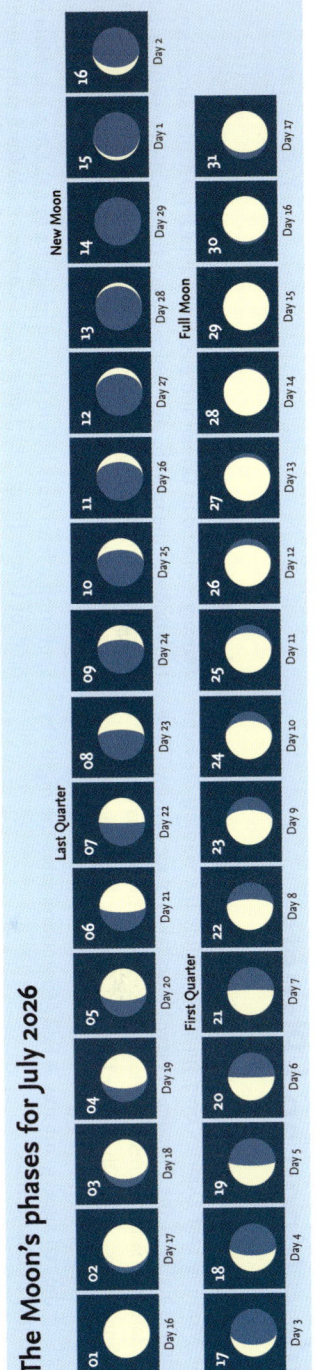

The Summer Triangle marked by the stars Deneb, Vega and Altair and the Milky Way.

The Moon's phases for July 2026

July – Moon and Planets

The Moon

The Moon is Last Quarter on 7 July. On 10 July the waning crescent Moon is 1.1°N of the **Pleiades**. A New Moon takes place 14 July. Three days later the waxing crescent moon passes 0.5°S of **Regulus** and 2.0°S of **Venus** (mag. -4.2). On 21 July the First Quarter Moon lies 2.4°S of **Spica**. On 24 July the waxing gibbous Moon approaches **Antares**, sitting 0.6°S of the red supergiant star. A Full Moon dominates the night sky on 29 July in **Capricornus**.

The Planets

Mercury is placed in **Gemini** and eventually appears just after sunset until the middle of the month, after which it leads the Sun at sunrise (mag. 2.2 to 5.6 then brightening again to 0.5). **Venus** stays in **Leo** for the month, appearing after sunset (mag. -4.1 to -4.3). On 9 July it passes 0.9°N of **Regulus**, visible until a few hours before midnight. **Mars** sits in **Taurus**, appearing a few hours before sunrise (mag. 1.4 to 1.3). On 4 July **Mars** is 0.1°S of **Uranus** (mag. 5.8). **Jupiter** lies in **Cancer**, visible from sunset until around 20:00. The gas giant approaches the Sun and leads the Sun in the dawn sky after 29 July (mag. -1.8). **Saturn** continues in **Pisces**, climbing above the eastern horizon after midnight (mag. 0.8 to 0.6). Saturn enters retrograde motion on 26 July at mag. 0.7. Uranus in Taurus is lost in the glare of the Sun at sunrise (mag. 5.8). **Neptune** in Pisces appears after midnight and enters retrograde motion on 7 July (mag. 7.8 to 7.7).

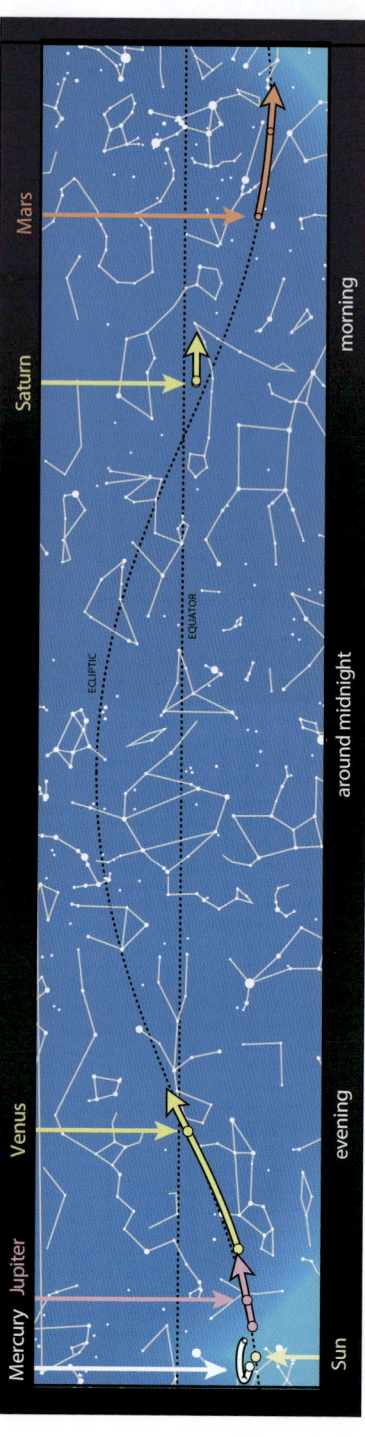

The path of the Sun and the planets along the ecliptic in July.

Calendar for July

06	17:30	Earth at aphelion (152,087,778 km = 1.01664 AU)
07	19:29	Last Quarter Moon
09	14:36	Venus (mag. -4.1) 0.9°N of Regulus
10	22:54	Pleiades 1.1°S of the Moon
13	07:50	Moon at perigee = 359,111 km
14	09:43	New Moon
17	00:07	Regulus 0.5°N of the Moon
17	16:31	Venus (mag. -4.2) 2.0°N of the Moon
21	03:21	Spica 2.4°N of the Moon
21	11:06	First Quarter Moon
24	21:00	Antares 0.6°N of the Moon
25	16:45	Moon at apogee = 405,549 km
28		Piscis Austrinid meteor shower maximum
29	14:36	Full Moon
30		Southern δ-Aquariid meteor shower maximum
30		α-Capricornid meteor shower maximum

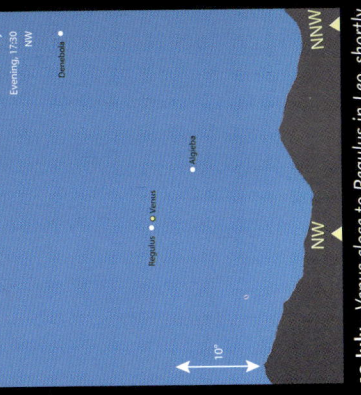

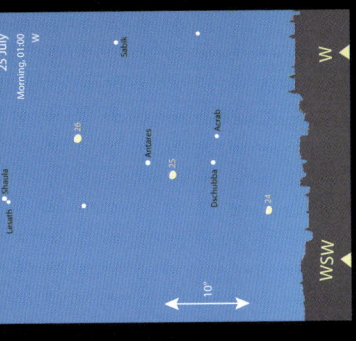

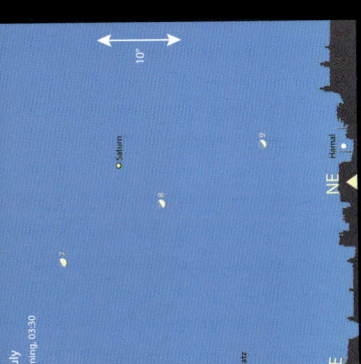

8 July • Last Quarter Moon in Pisces with Saturn, rising in the east after midnight.

10 July • Venus close to Regulus in Leo, shortly after sunset.

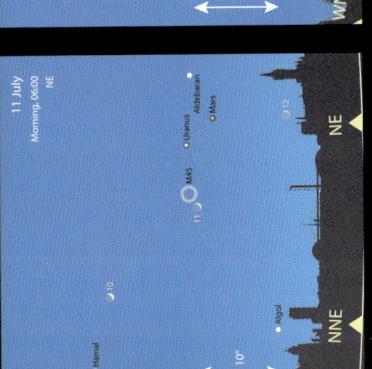

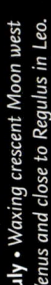

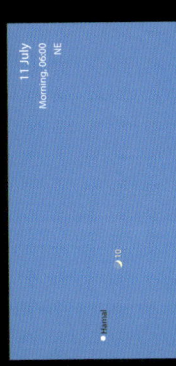

11 July • The waning crescent Moon close to the Pleiades, Uranus and Mars. Aldebaran and Elnath (β Tau) are nearby.

17 July • Waxing crescent Moon west of Venus and close to Regulus in Leo.

25 July • Waxing gibbous Moon south of Antares. Sabik (η Oph) and Saik (ζ Oph) in Ophiuchus sit westwards.

JULY 75

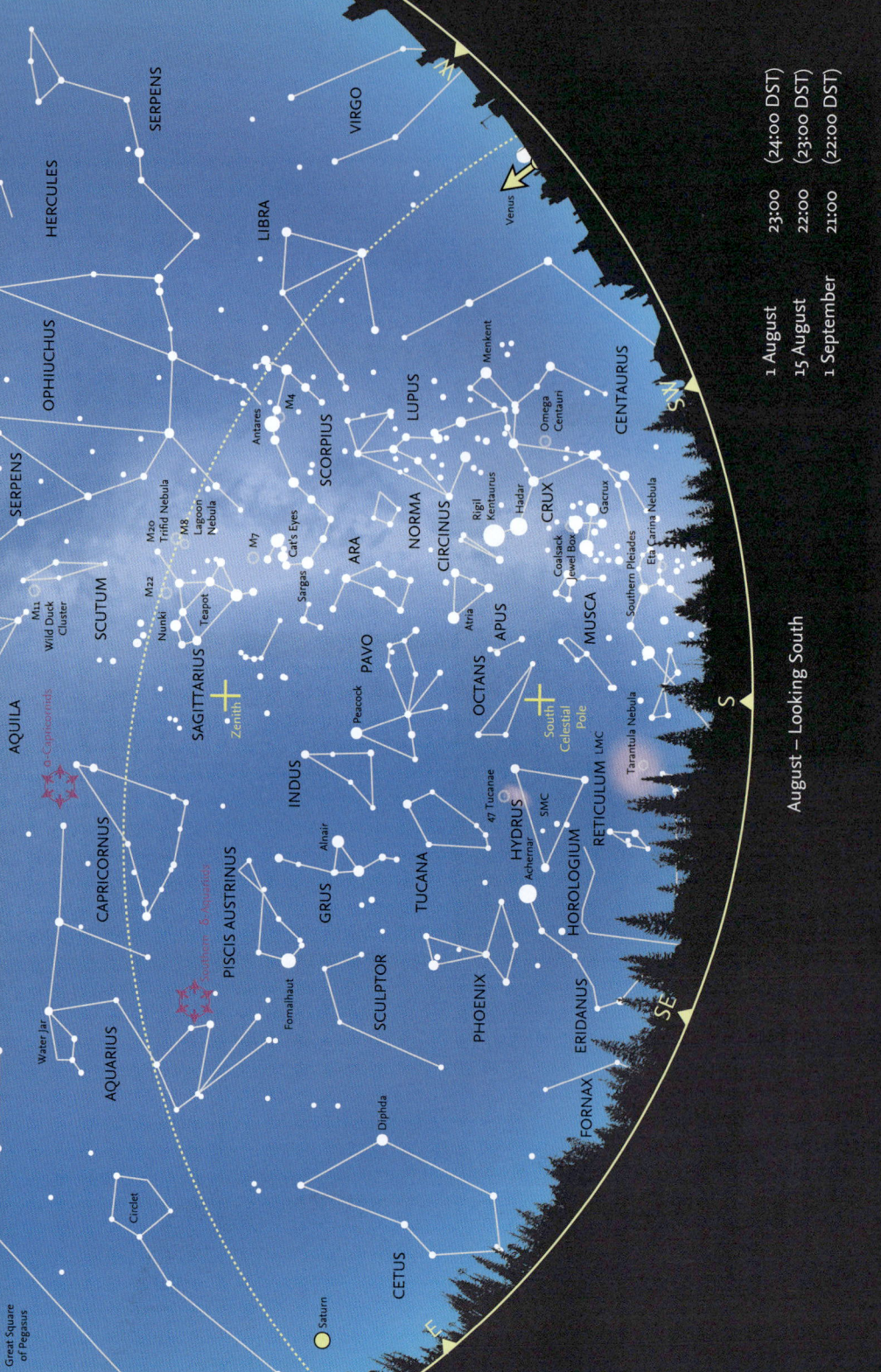

August – Looking South

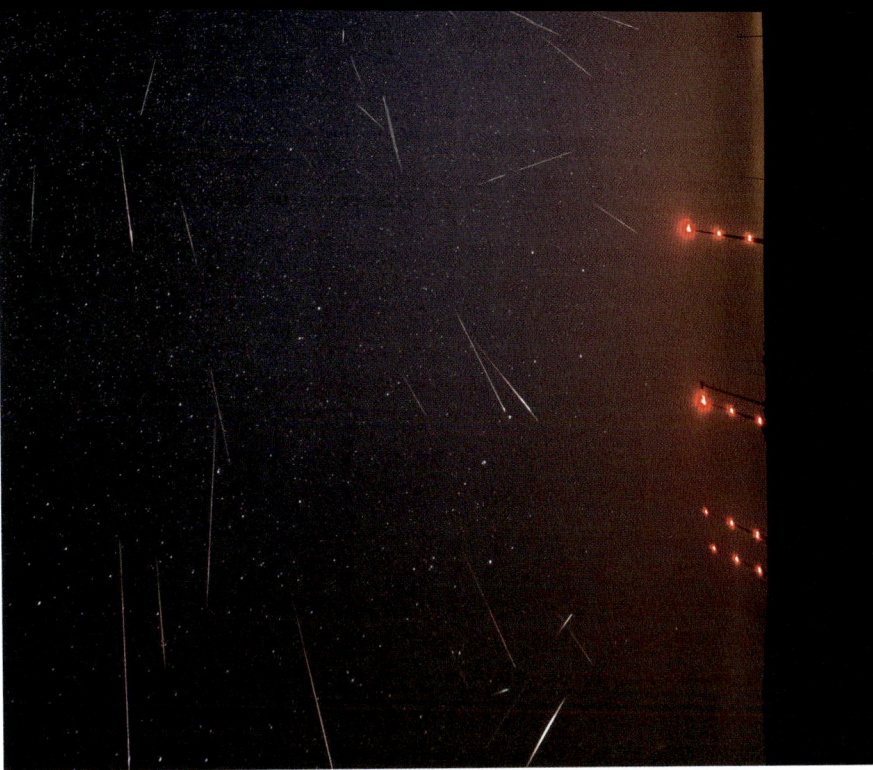

The Perseid meteor shower – a compilation of images.

Three constellations: ***Volans***, ***Chamaeleon*** and ***Octans*** are on the meridian, with ***Pavo*** (the Peacock) higher towards the zenith. Much of ***Carina*** is below the horizon, although the ***Eta Carinae Nebula*** and the ***Southern Pleiades*** are still visible. The ***Large Magellanic Cloud*** (LMC) and the faint constellation of ***Mensa*** are slightly higher in the sky. ***Crux*** is now lower, but the whole of ***Centaurus*** and ***Lupus*** remains visible. Much of ***Virgo*** has set in the west and ***Libra*** is following it down towards the horizon. ***Scorpius*** and ***Sagittarius*** are still visible high in the sky. ***Achernar***, the ***Small Magellanic Cloud*** (SMC) and ***Hydrus*** are now well clear of the horizon, but below them are more small constellations: ***Dorado***, ***Reticulum*** and ***Horologium***. More of the river of ***Eridanus*** is visible, together with parts of ***Pictor*** and ***Fornax***. Much of ***Cetus*** has risen and the whole of the western arm of ***Pisces*** is now clearly seen. ***Sculptor***, ***Grus*** and ***Piscis Austrinus*** lie halfway between the eastern horizon and the zenith, with faint eighteenth-century constellation ***Microscopium*** closer to the actual zenith.

Meteors

The ***Piscis Austrinid*** shower continues until about 10 August, after maximum on 28 July. The ***Southern δ-Aquariids*** reach maximum of up to 25 meteors per hour on 30 July. The ***Perseids*** peak on 13 August, when the rate may reach as high as 100 meteors per hour. These are best seen shortly before sunrise very close to the horizon. In 2026, there is a New Moon at the maximum, so there will be no competition from moonlight. The Perseids are debris from Comet 109P/Swift-Tuttle (the Great Comet of 1862). Perseid meteors are fast and many of the brighter ones leave persistent trains. Some bright fireballs also occur during the shower, arising from larger chunks of cometary debris. The ***α-Aurigid*** shower begins on 28 August and is a short shower, lasting until 5 September and reaching maximum in 2026 on 1 September.

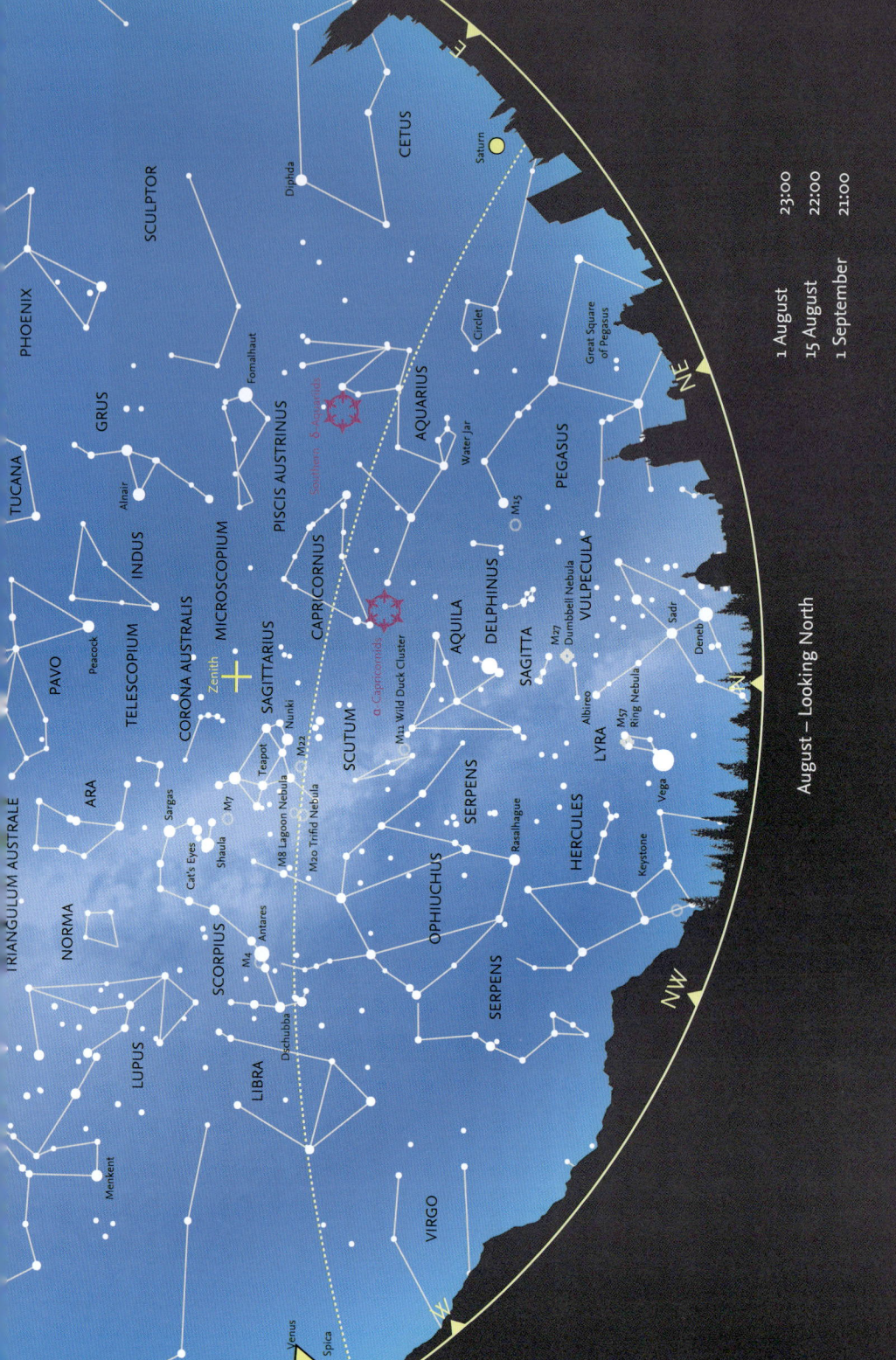

August – Looking North

Cygnus, with brilliant **Deneb** (α Cygni), is now prominent just to the east of the meridian. The star **Albireo** (β Cygni) marks the Swan's beak and through a telescope it can be seen to be an optical double: the brighter star is an orange giant 100 times more luminous than the Sun; the second star is blue-white and 230 times more luminous than the Sun. The stars are not thought to be orbiting each other (in a binary system). The other two stars of the (northern) Summer Triangle, **Vega** in **Lyra** and **Altair** in **Aquila** are also unmistakable in the sky. West of Aquila, mainly in the star clouds of the Milky Way, lies **Scutum**, most famous for the bright open cluster, **M11** or the Wild Duck Cluster, readily visible in binoculars. **Hercules**, however, is beginning to descend towards the northwestern horizon. Above it, the sprawling constellation of **Ophiuchus** is readily seen. The Great Square of **Pegasus** is now visible in the east, although **Alpheratz** (α Andromedae) is low on the horizon. The western side of **Pisces**, as well as **Aquarius** and **Capricornus**, is fully visible along the ecliptic. **Sagittarius** contains many gaseous nebulae, like **M8** (the star-forming Lagoon Nebula, close to γ Sagittarii) and **M20** (the Trifid Nebula) and several globular clusters such as **M22**. The constellation is at the zenith, with Scorpius to the west.

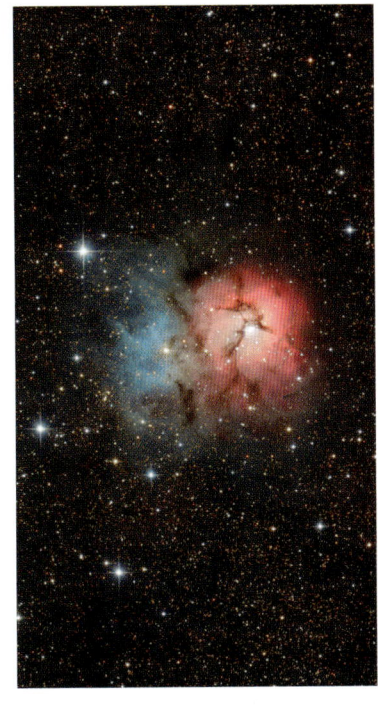

The Trifid Nebula (M20), a hydrogen emission cloud in a star-forming region of the Milky Way.

The Moon's phases for August 2026

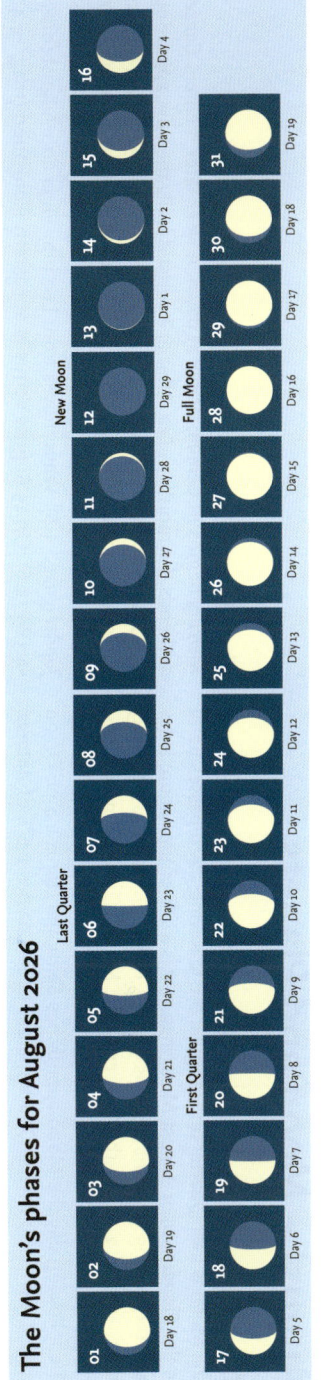

AUGUST

August – Moon and Planets

The Moon

The Moon is Last Quarter on 6 August; the following day it passes 1.2°N of the *Pleiades*. On 9 August *Mars* (mag. 1.3) is 4.4°S of the Moon; a day later the waning crescent Moon is 3.6°S of *Pollux*. On 11 August *Mercury* (mag. -0.9) is 2.1°S of the thin crescent Moon. On 16 August *Venus* (mag. -4.4) lies 2.1°N of the waxing crescent Moon in the early evening. The following day the Moon approaches *Spica*, the star is 2.4°N. On 20 August the Moon is First Quarter; a day later it passes 0.6°S of *Antares*. On 28 August a full or partial lunar eclipse will be visible from some regions in the southern hemisphere such as Brazil and New Zealand, maximum coverage occurs at 04:13 UT.

The Planets

Mercury begins in *Gemini*, visible at dawn. A week later it enters *Cancer* and ends the month in *Leo*. After 27 August it trails the Sun, appearing shortly after sunset. Its magnitude brightens from 0.3 to -1.9 (at solar conjunction) and then dims to -1.6. It reaches western elongation on 2 August (mag. 0.2). On 15 August Mercury (mag. -1.2) is 0.5°N of *Jupiter* (mag. -1.8). *Venus* moves from Leo into *Virgo* on 1 August (mag. -4.3 to -4.6). On 15 August Venus is at eastern elongation (mag. -4.4). *Mars* starts in *Taurus* and swings into Gemini in the middle of the month; it brightens from 1.4 to 1.2 and dims slightly to 1.3 at the end of the month. Mars appears 1.5°N of (**1**) *Ceres* on 9 August in Taurus. Jupiter can be seen just before sunrise in Cancer (mag. -1.8). *Saturn* is in *Pisces*, visible after 23:00 (mag. 0.6 to 0.5). *Uranus* lies in Taurus, above the horizon from around midnight (mag. 5.8 to 5.7). *Neptune* is in Pisces and visible after 22:00 (mag. 7.7).

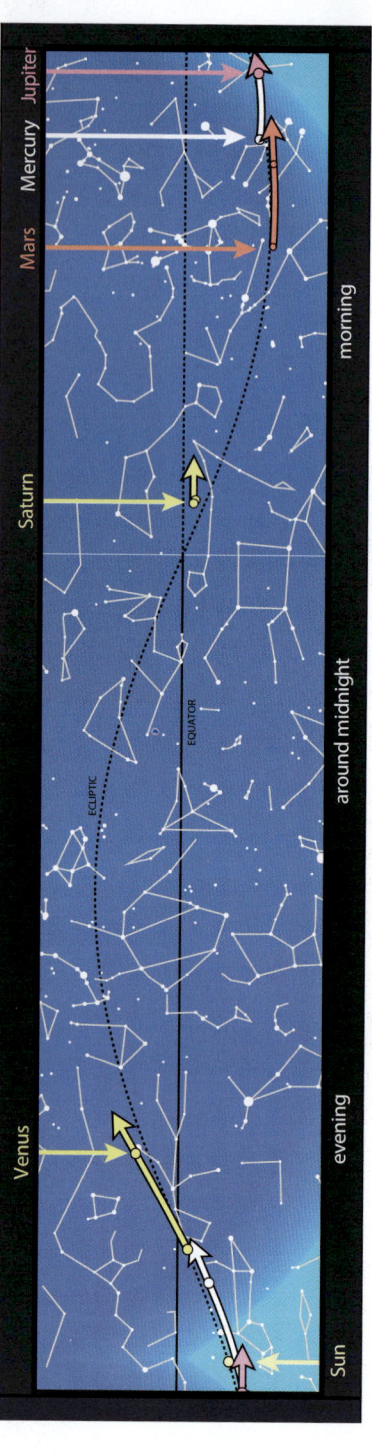

The path of the Sun and the planets along the ecliptic in August.

Calendar for August

02	08:09	Mercury at western elongation (19.5°W, mag. 0.2)
06	02:21	Last Quarter Moon
07	06:23	Pleiades 1.2°S of the Moon
09	05:31	Mars (mag. 1.3) 4.4°S of the Moon
10	11:18	Moon at perigee = 363,288 km
10	22:38	Pollux 3.6°N of the Moon
11	12:48	Mercury (mag. -0.9) 2.1°S of the Moon
12	17:37	New Moon
12	17:46	Total solar eclipse (Iceland, Spain, Greenland and the Arctic). Partial eclipse visible from North America, western Africa & Europe
12–13		Perseid meteor shower maximum
15	06:00	Venus at eastern elongation (45.9°E, mag. -4.4)
16	08:47	Venus (mag. -4.4) 2.1°N of the Moon
17	11:49	Spica 2.4°N of the Moon
20	02:46	First Quarter Moon
21	04:18	Antares 0.6°N of the Moon. An occultation will be visible from Antarctica, southern Argentina, Chile & Falkland Islands
22	08:20	Moon at apogee = 404,644 km
28	04:13	Partial lunar eclipse. Visible from Europe, Africa & the eastern Pacific
28	04:18	Full Moon

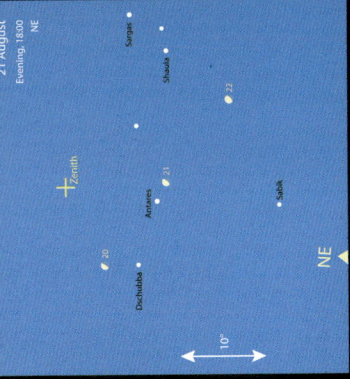

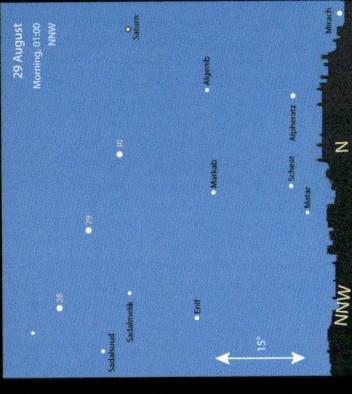

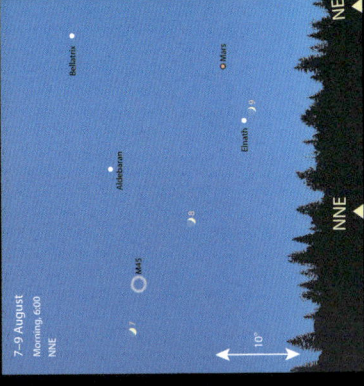

7–9 August • *The waning Moon after 6 August, moving past the Pleiades and Mars as a waning crescent.*

21 August • *First Quarter Moon close to Antares in Scorpius.*

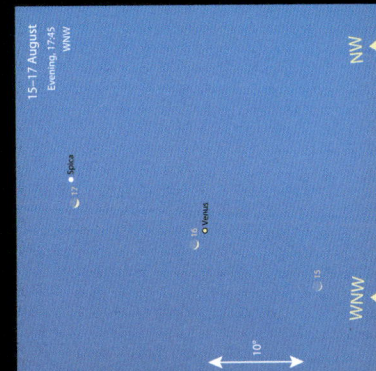

15–17 August • *The waxing crescent Moon and Venus at greatest eastern elongation on 15 August. Moon next to Spica on 17 August.*

29 August • *Full Moon in Aquarius, close to Sadelmelik (α Aquarii). Saturn due north.*

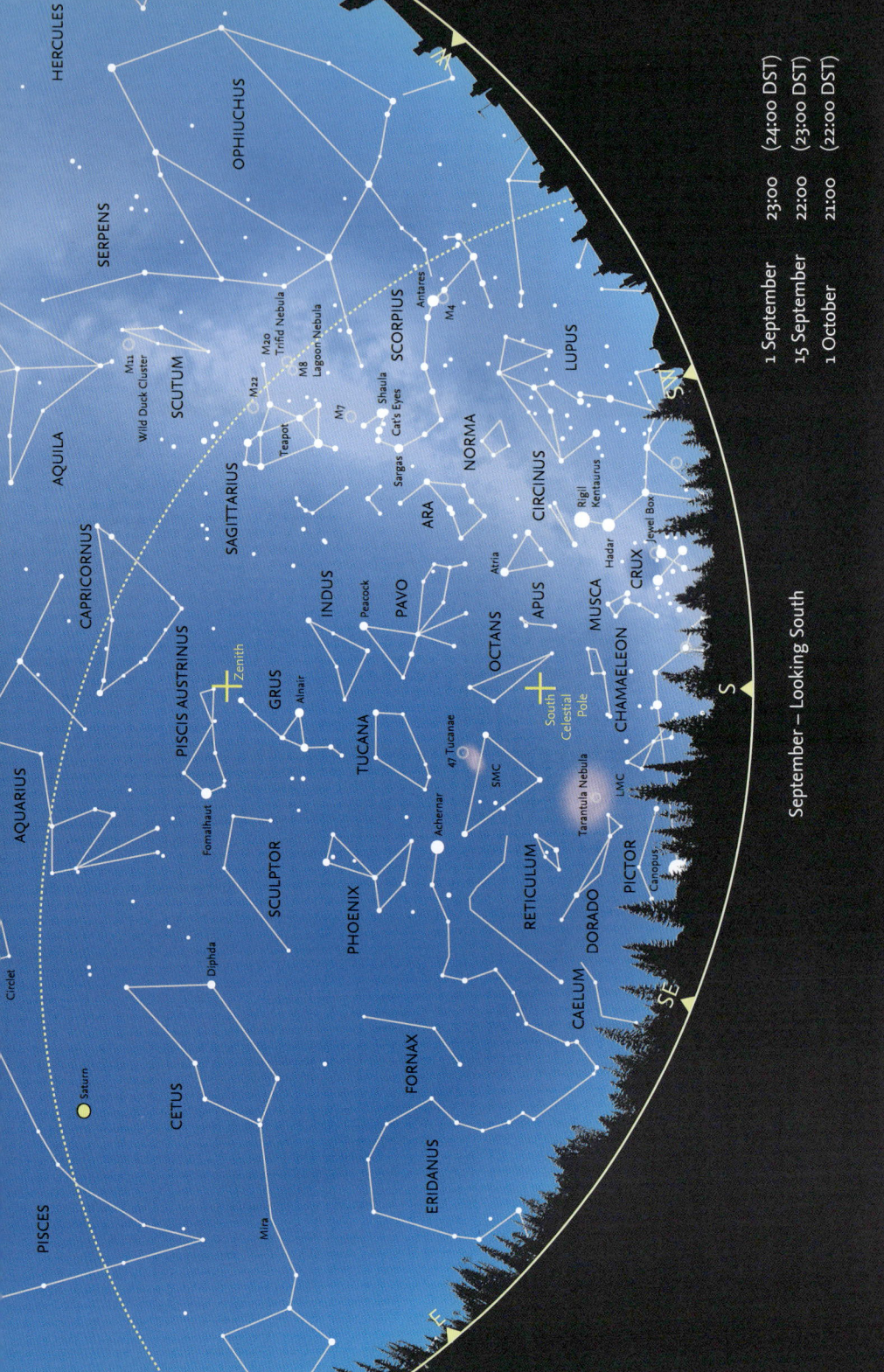

September – Looking South

The southern vernal (spring) equinox occurs on 23 September, when the Sun moves south of the equator in *Virgo*.

The smallest of the constellations, **Crux**, is now very low, but **Canopus** (α Carinae), the second brightest star in the sky, has now become visible once more. **Dorado** and the **Large Magellanic Cloud** (LMC) are higher and easier to see, as are the faint constellations of **Pictor**, **Caelum** and **Reticulum**. The **Small Magellanic Cloud** (SMC) and **47 Tucanae** (the globular cluster in *Tucana*) are now nearly halfway between the horizon and the zenith. Although becoming low, the whole of **Centaurus** and **Lupus** remain visible in the southwest. **Scorpius**, with red-orange *Antares*, is beginning to descend in the west, but **Sagittarius** is clearly seen high in the sky. The constellation of **Piscis Austrinus**, with **Fomalhaut** (α Piscis Austrini), is at the zenith. Below it are the constellations of **Grus** and **Pavo**, with **Achernar** (α Eridani) and almost the whole of the long river constellation of **Eridanus**.

Meteors

There are two minor showers in September. A few meteors may be seen from the α-**Aurigid** shower, active from late August, with a maximum on 1 September. In 2026, the Moon will be a bright waning gibbous, so conditions are unfavourable during the shower. At maximum, however, the hourly rate hardly reaches 10 meteors per hour, although the meteors are bright and relatively easy to photograph. The **Southern Taurid** shower begins this month (on 10 September) and, although rates are low (less than six per hour), often produces very bright fireballs. This is a very long shower, lasting until about 20 November.

Fomalhaut in the upper left, the brightest star in the constellation of Piscis Austrinus, the Southern Fish.

SEPTEMBER 83

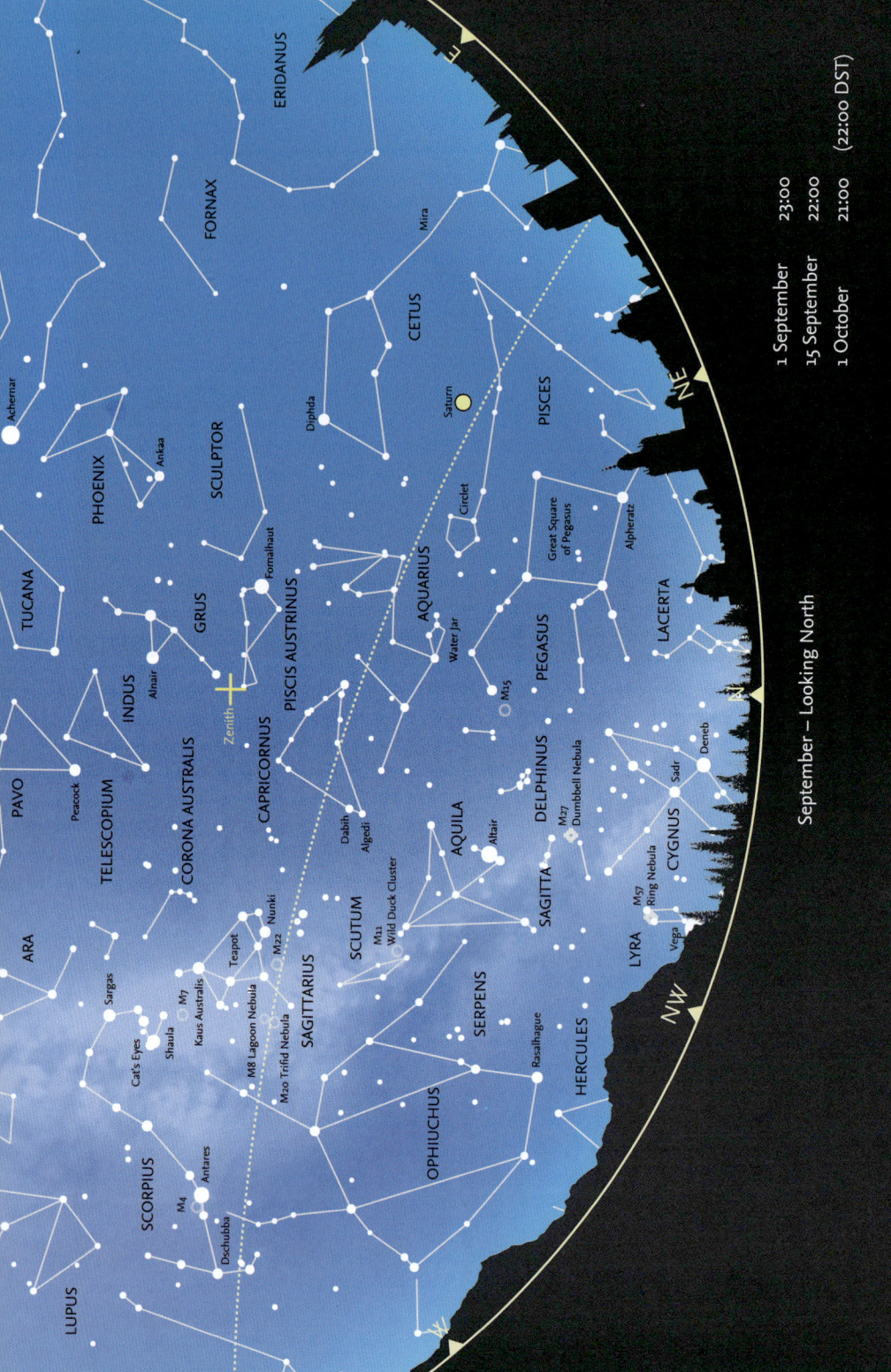

September – Looking North

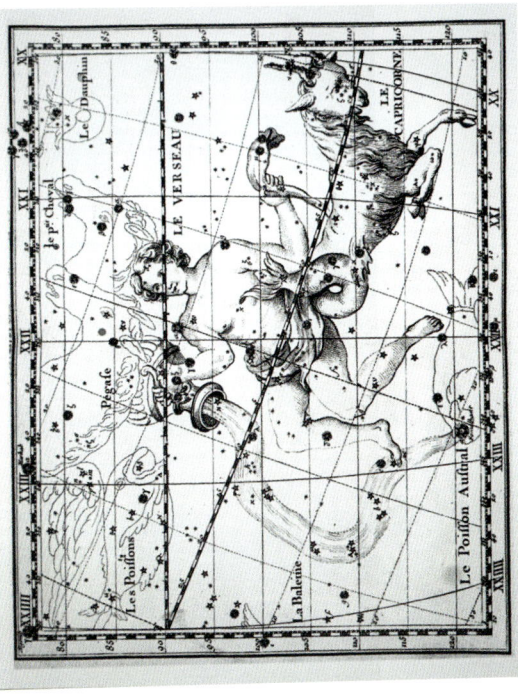

The Great Square of **Pegasus** is nearing the meridian from the east and, on the west, the three bright stars of the (northern) Summer Triangle, **Deneb** (α Cygni), **Altair** (α Aquilae) and **Vega** (α Lyrae) are still clearly seen, although Vega is low on the horizon, as is **M31** the spectacular galaxy in **Andromeda**. **Delphinus** is well placed, as are the fainter constellations of **Sagitta** and **Scutum** along the Milky Way. Giant **Ophiuchus** is beginning to descend towards the western horizon and much of **Libra** has disappeared. The three zodiacal constellations of **Pisces**, **Aquarius** and **Capricornus** are readily visible. **Algedi** (α Capricorni) is actually an optical double with the two stars (α¹ Cap and α² Cap) readily seen with the naked eye. **Dabih** (β Capricorni), just to the south, is also a double star and the components are relatively easy to separate with binoculars. In Aquarius, there is a small asterism consisting of four stars, resembling a tiny letter 'Y', known as the Water Jar. Somewhat higher, Piscis Austrinus is at the zenith. In classical illustrations, water is shown flowing from the Water Jar towards bright **Fomalhaut** (α Piscis Austrini). **Scorpius** and **Sagittarius** are now on the western side of the sky.

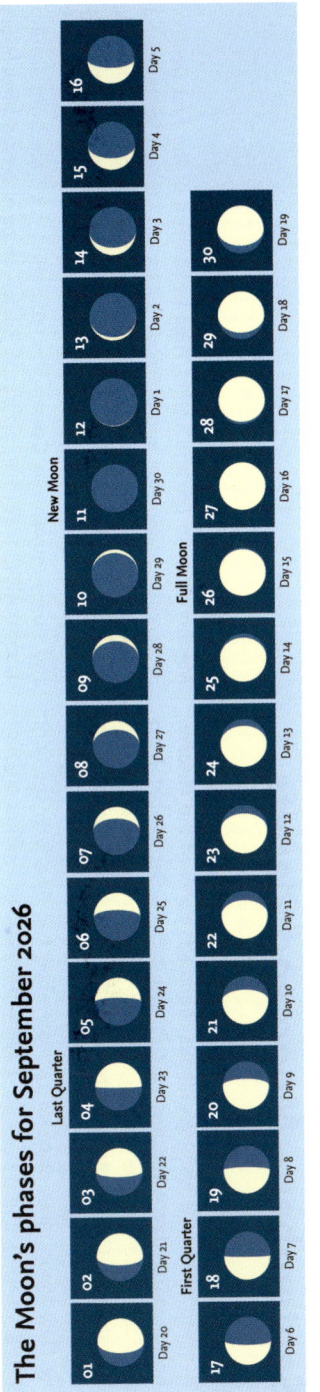

Aquarius including the Water Jar asterism and Fomalhaut (α Piscis Austrini).

SEPTEMBER

September – Moon and Planets

The Moon

On 3 September the waning gibbous Moon sits 1.2°N of the *Pleiades*. The following day it reaches Last Quarter. On 6 September *Mars* (mag. 1.2) is 3.0°S of the waning crescent Moon; a day later it moves 3.6°S of *Pollux*. On 8 September the Moon moves 0.8°N of *Jupiter* (mag. -1.8); the following day the thin crescent Moon is 0.5°S of *Regulus*. The New Moon takes place on 11 September; two days later the faint waxing crescent Moon is 2.4°S of *Spica*. *Venus* (mag. -4.8) is 0.5°S of the Moon on 14 September; an occultation will occur in the morning, lost in the daylight. On 17 September *Antares* lies 0.6°N of the waxing crescent Moon, followed by a First Quarter Moon a day later. A Full Moon occurs on 26 September and the Moon rejoins the Pleiades on 30 September, lying 1.1°N of the star cluster.

The Planets

Mercury starts in *Leo* and enters *Virgo* by the end of the first week of September, visible after sunset (mag. -1.5 to -0.1). On 26 September it moves 0.8°N of *Spica*, at mag. -0.2. *Venus* sits in Virgo for the month, approaching the Sun and visible in the early evening (mag. -4.6 to -4.8). On 1 September it is 1.2°S of Spica. *Mars* starts the month in *Gemini* and slowly moves into *Cancer*. The red planet is visible from around 02:00 until sunrise (mag. 1.2 to 1.1). *Jupiter* can be observed a few hours before sunrise in Cancer, crossing into Leo near the end of the month (mag. -1.8 to -1.9). *Saturn* lies on the border between *Pisces* and *Cetus*, visible after 21:00 (mag. 0.5 to 0.3). *Uranus* stays in *Taurus*, visible in the late evening (mag. 5.7). It enters retrograde motion on 10 September. *Neptune* is in Pisces (mag. 7.7), reaching opposition on 26 September, a distance of 28.9 AU from Earth.

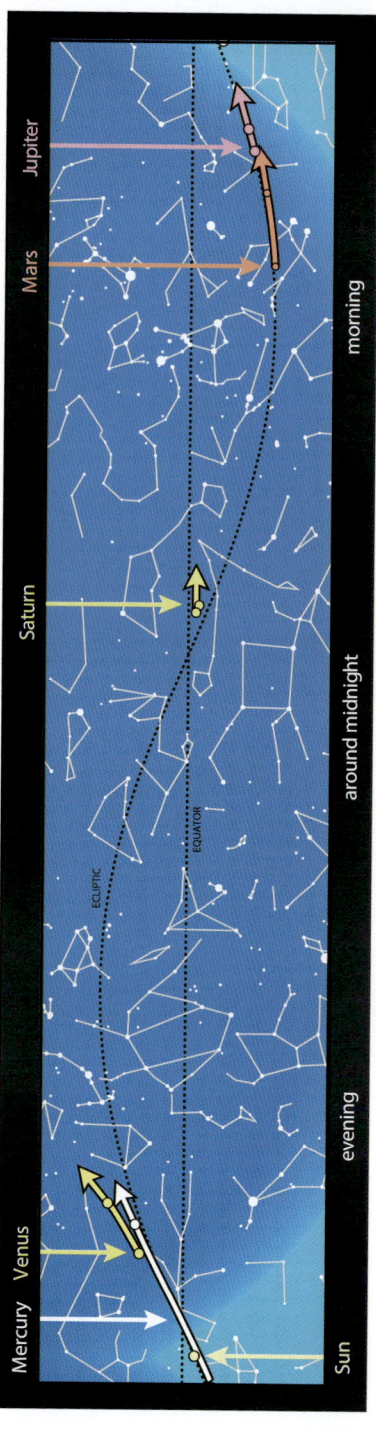

The path of the Sun and the planets along the ecliptic in September.

Calendar for September

01		α-Aurigid meteor shower maximum
01	13:24	Venus (mag. -4.6) 1.2°S of Spica
03	12:03	Pleiades 1.2°S of the Moon
04	07:51	Last Quarter Moon
06	18:24	Mars (mag. 1.2) 3.0°S of the Moon
06	20:26	Moon at perigee = 368,255 km
07	06:32	Pollux 3.6°N of the Moon
08	18:13	Jupiter (mag. -1.8) 0.8°S of the Moon
09	19:36	Regulus 0.5°N of the Moon
11	03:27	New Moon
13	20:53	Spica 2.4°N of the Moon
14	11:10	Venus (mag. -4.8) 0.5°S of the Moon
17	12:18	Antares 0.6°N of the Moon. Occultation visible from Antarctica, Australia & Tasmania
18	20:44	First Quarter Moon
19	03:00	Moon at apogee = 404,217 km
23	00:06	Autumnal Equinox
26	01:49	Mercury (mag. -0.2) 0.8°N of Spica
26	16:49	Full Moon
30	17:39	Pleiades 1.1°S of the Moon

4 September • *Waning gibbous Moon next to the Pleiades, Uranus nearby, along with Aldebaran, Capella, Elnath and Betelgeuse.*

7 September • *Waning crescent Moon approaching Mars in Gemini, Pollux and Castor close by.*

9 September • *Waning crescent Moon and Jupiter, Mars to the north, Asellus Australis (δ Cnc) in Cancer nearby.*

14 September • *Waxing crescent Moon and Venus just after sunset. Mercury low in the west.*

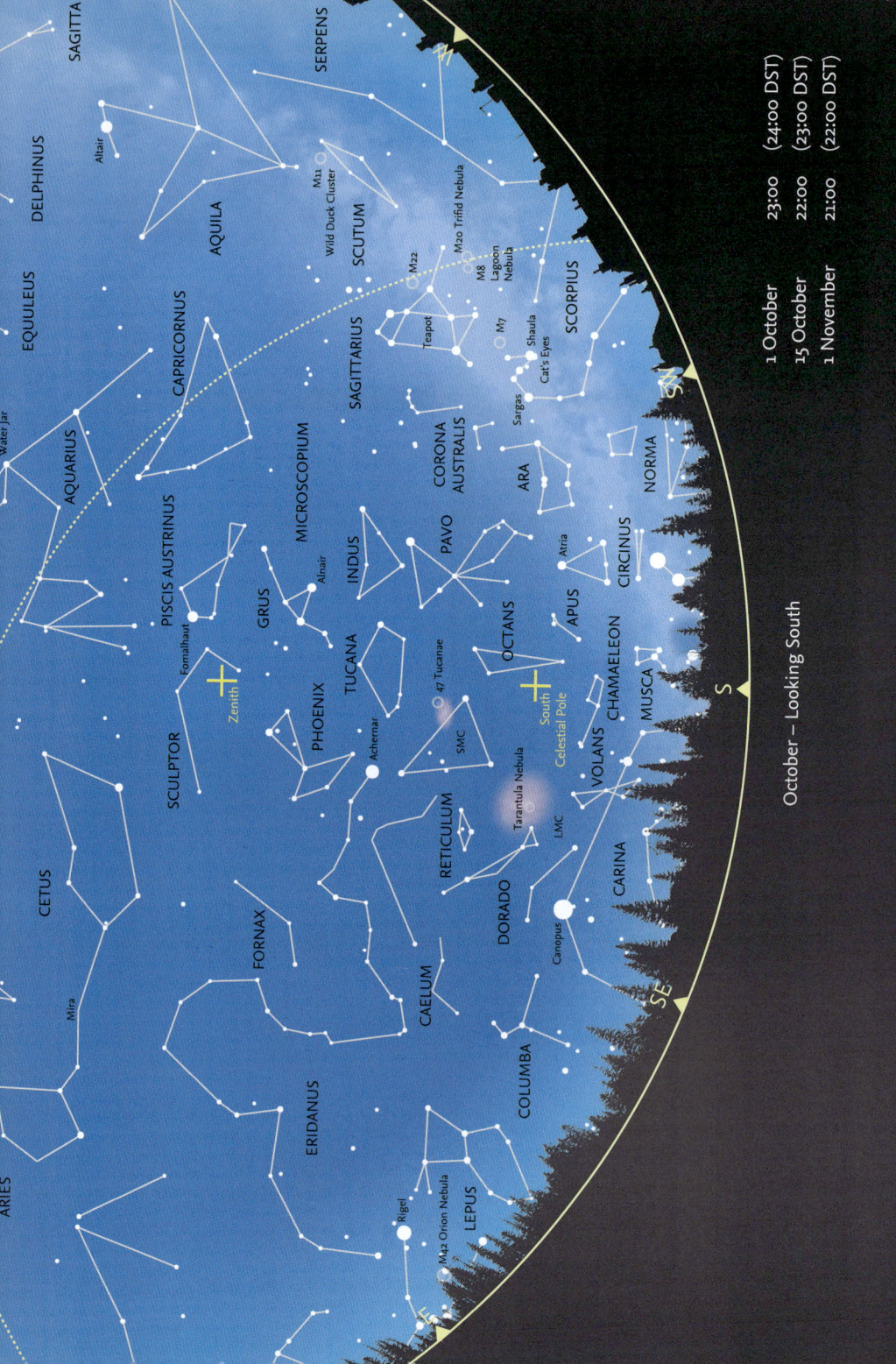

October – Looking South

Daylight Saving Time starts where used in Australia and in New Zealand on Sunday 4 October. **Crux** is now extremely low, only just visible above the horizon. The **False Cross** has reappeared, also very low, farther towards the east on the opposite side of the meridian. **Canopus** (α Carinae) is now much higher, as is the **Large Magellanic Cloud** (LMC). The **Small Magellanic Cloud** (SMC) and the globular cluster **47 Tucanae** are on the southern meridian, halfway to the zenith, as is **Achernar** (α Eridani). The whole of the long constellation of **Eridanus** is now visible together with **Rigel** in **Orion**. Still higher are **Phoenix**, **Grus** and **Piscis Austrinus**. **Pavo** has begun to descend in the southwest. **Ophiuchus** has now slipped below the horizon as has much of **Scorpius**, only the 'tail' of which remains visible. **Sagittarius** is getting lower, but remains visible, as do the zodiacal constellations of **Capricornus** and **Aquarius**.

Meteors

The **Orionids** meteor shower starts on 2 October and builds up to a broad maximum lasting around a week centred on 21 October. Like the May η-**Aquariid** shower, the Orionids are associated with Comet 1P/Halley. During this second pass through the stream of particles from the comet, slightly fewer meteors are seen than in May. In both showers the meteors are very fast and many leave persistent trains with hourly rates around 15. In 2026, there is a waxing gibbous Moon lighting up the sky and interfering with observations. Occasionally, rates are higher (50–70 per hour). The faint shower of the **Southern Taurids** (often with bright fireballs) peaks on 10 October when the Moon is New and skies are dark. Towards the end of the month, another shower (the **Northern Taurids**) begins to show activity, peaking in November. The parent comet for both Taurid showers is Comet 2P/Encke. The meteors in both Taurid streams are relatively slow and bright.

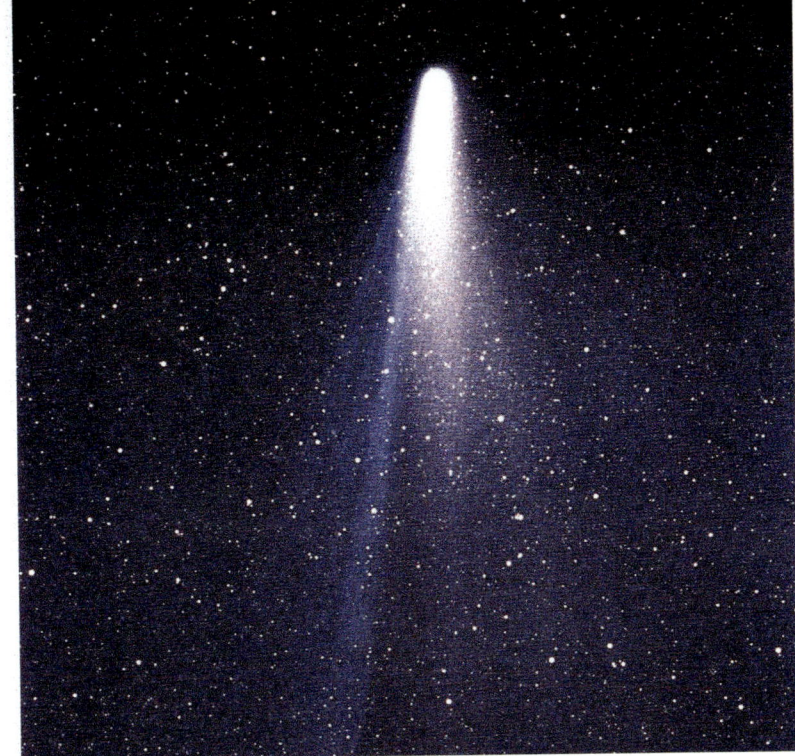

Comet 1P/Halley imaged in March 1986. Debris from the comet provide the source of the Orionids meteor shower.

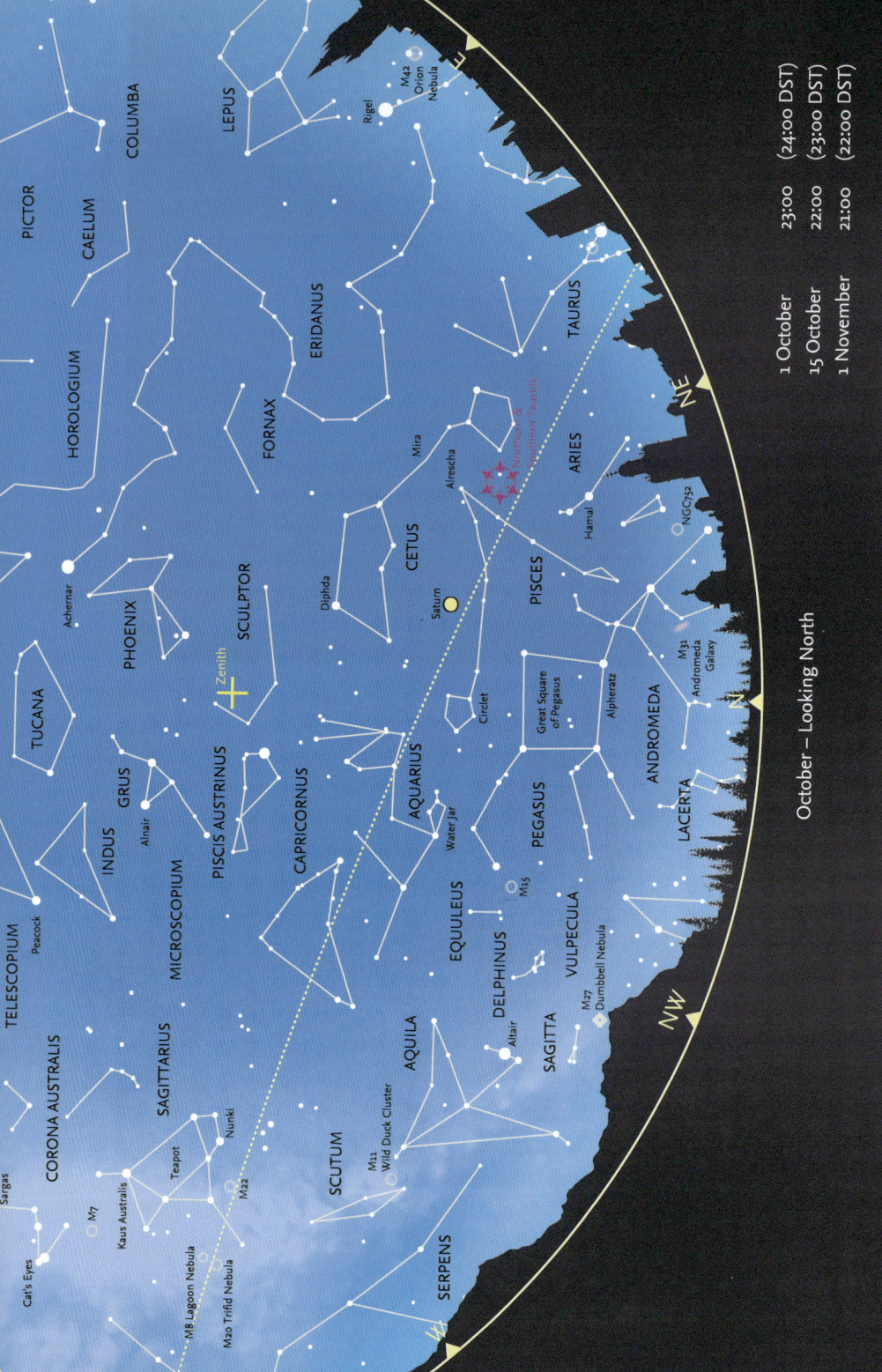

October – Looking North

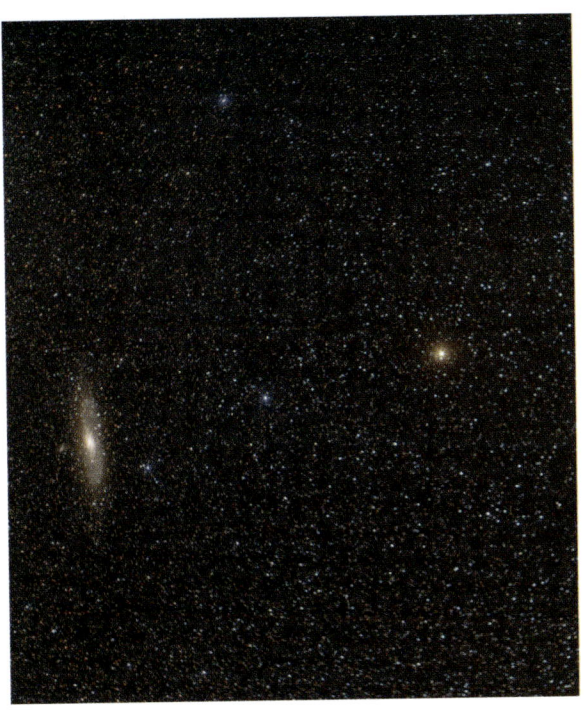

The Great Square of *Pegasus* now dominates the sky to the north and the whole of *Andromeda* with the *Andromeda Galaxy* (M31) is now clear of the horizon. The two chains of stars forming *Pisces* frame the Great Square together with *Alrescha* (α Piscium) at the point where the two lines of stars join. Farther along the ecliptic, the constellations of *Aquarius* and *Capricornus* are fully visible. Farther east, all of *Cetus* is visible and, low down on the horizon, *Aldebaran* in *Taurus* is beginning to rise. Due east, *Orion* is becoming visible, with blue-white *Rigel* and the beginning of the long, winding constellation of *Eridanus* that winds its way to *Achernar* (α Eridani), far to the south. West of the meridian, *Lyra*, with *Vega* has disappeared; much of *Cygnus* is invisible and *Deneb* is brushing the horizon. Only *Aquila* and *Altair* (α Aquilae) remain clearly visible. *Vulpecula*, with the planetary nebula M27 is still visible – as is *Sagitta* – and above them, the distinctive constellation of *Delphinus* and the very faint constellation of *Equuleus* (the Little Horse).

Andromeda galaxy next to the stars ν Andromedae, μ Andromedae and Mirach (β Andromedae).

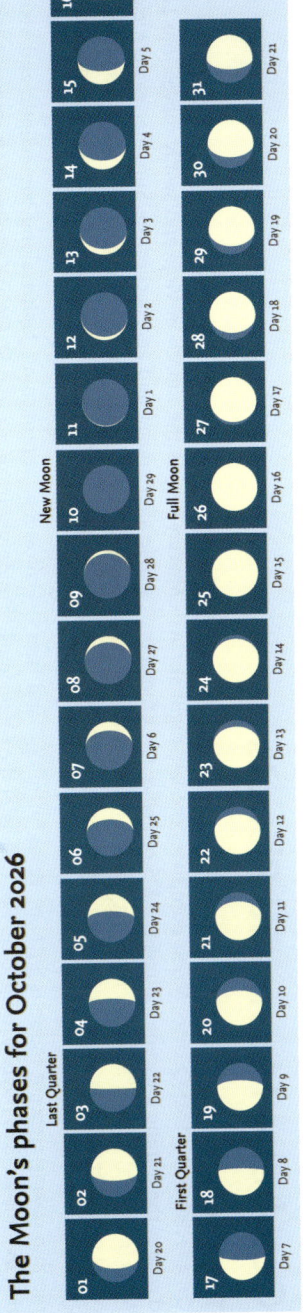

The Moon's phases for October 2026

October – Moon and Planets

The Moon

On 3 October the Last Quarter Moon dominates the late night sky. A day later the waning crescent Moon sits 3.8°S of **Pollux**. On 5 October **Mars** (mag. 1.1) sits 1.2°S of the Moon; a day later the Moon moves 0.2°N of **Jupiter** (mag. -1.9). On 7 October **Regulus** is 0.6°N of the faint crescent Moon; a New Moon takes place on 10 October. On 12 October the thin waxing crescent Moon is 3.1°N of **Venus** (mag. -4.5) and later that day it is 2.1°S of **Mercury** (mag. 0.0). Two days later the Moon joins **Antares**, sitting 0.4°S of the red star. On 18 October the Moon is First Quarter; a Full Moon occurs on 26 October. Two days later it lies 1.0°N of the **Pleiades** in **Taurus**. The waning gibbous Moon rejoins **Pollux** on 31 October, sitting 4.0°S of one of the heads of **Gemini**.

The Planets

Mercury begins the month in **Virgo** and swiftly moves into **Libra**. It is visible after sunset and approaches the Sun after eastern elongation on 12 October (mag. -0.1 to 2.6). **Venus** is in Virgo visible in the early evening. It approaches the Sun and is seen at dawn after 24 October (mag. -4.8 to -4.1). **Mars** is in **Cancer**, it progresses into **Leo** for Halloween, approaching **Jupiter** along the ecliptic and visible after midnight (mag. 1.1 to 0.8). On 11 October Mars (mag. 1.1) sits next to the Beehive Cluster, **M44** (mag. 3.1), 1.0°N. Jupiter is in Leo, climbing above the horizon from 01:00 (mag. -1.9 to -2.0). **Saturn** lies on the border between **Pisces** and **Cetus**, visible after sunset (mag. 0.3). On 4 October it is at opposition (mag. 0.3), positioned 8.4 AU from Earth. **Uranus** is in **Taurus** and visible in the late evening (mag. 5.7 to 5.6). **Neptune** is in Pisces and visible just after sunset (mag. 7.7).

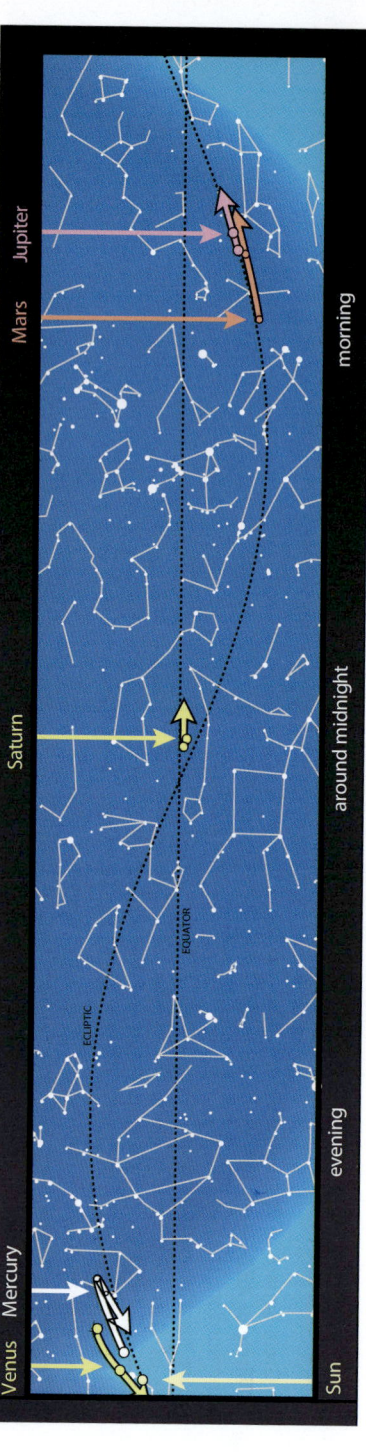

The path of the Sun and the planets along the ecliptic in October.

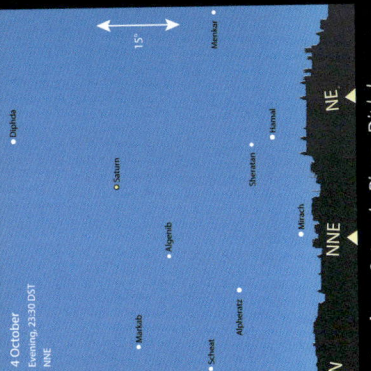

4 October
Evening, 23:30 DST
NNE

4 October
Morning, 04:30 DST
NE

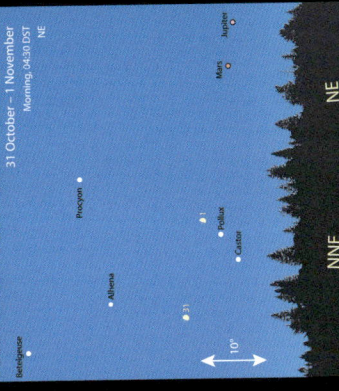

4 October
Morning, 04:30 DST
NE

31 October – 1 November
Morning, 04:30 DST
NE

4 October • Saturn in Pisces. Diphda, Alpheratz, Algenib and Markab nearby.

4 October • Last Quarter Moon in Gemini with Mars in adjacent Cancer. Pollux, Castor and Elnath lie close by.

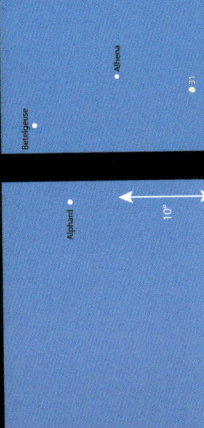

6 October
Morning, 05:30 DST
NE

6 October • Waning crescent Moon next to Mars, Jupiter sits nearby, Regulus in the vicinity.

31 October – 1 November • Waning gibbous Moon close to Pollux and Castor. Mars and Jupiter rising in the east.

Calendar for October

01	20:41	Moon at perigee = 369,338 km
03	13:25	Last Quarter Moon
04	12:21	Saturn (mag. 0.3) at opposition
04	12:27	Pollux 3.8°N of the Moon
05	05:30	Mars (mag. 1.1) 1.2°S of the Moon
06	10:18	Jupiter (mag. -1.9) 0.2°S of the Moon
07	02:57	Regulus 0.6°N of the Moon. Occultation visible from Africa
10	15:50	New Moon
10		Southern Taurid meteor shower maximum
12	02:30	Venus (mag. -4.5) 3.1°S of the Moon
12	10:00	Mercury at eastern elongation (25.2°E, mag. 0.0)
12	20:08	Mercury (mag. 0.0) 2.1°N of the Moon
14	20:25	Antares 0.4°N of the Moon. Occultation visible from eastern Uruguay, southern Brazil, South Georgia, the South Sandwich Islands & Saint Helena
16	22:56	Moon at apogee = 404,639 km
18	16:13	First Quarter Moon
21–22		Orionid meteor shower maximum
26	04:12	Full Moon
28	01:11	Pleiades 1.0°S of the Moon
28	18:01	Moon at perigee = 364,411 km
31	18:00	Pollux 4.0°N of the Moon

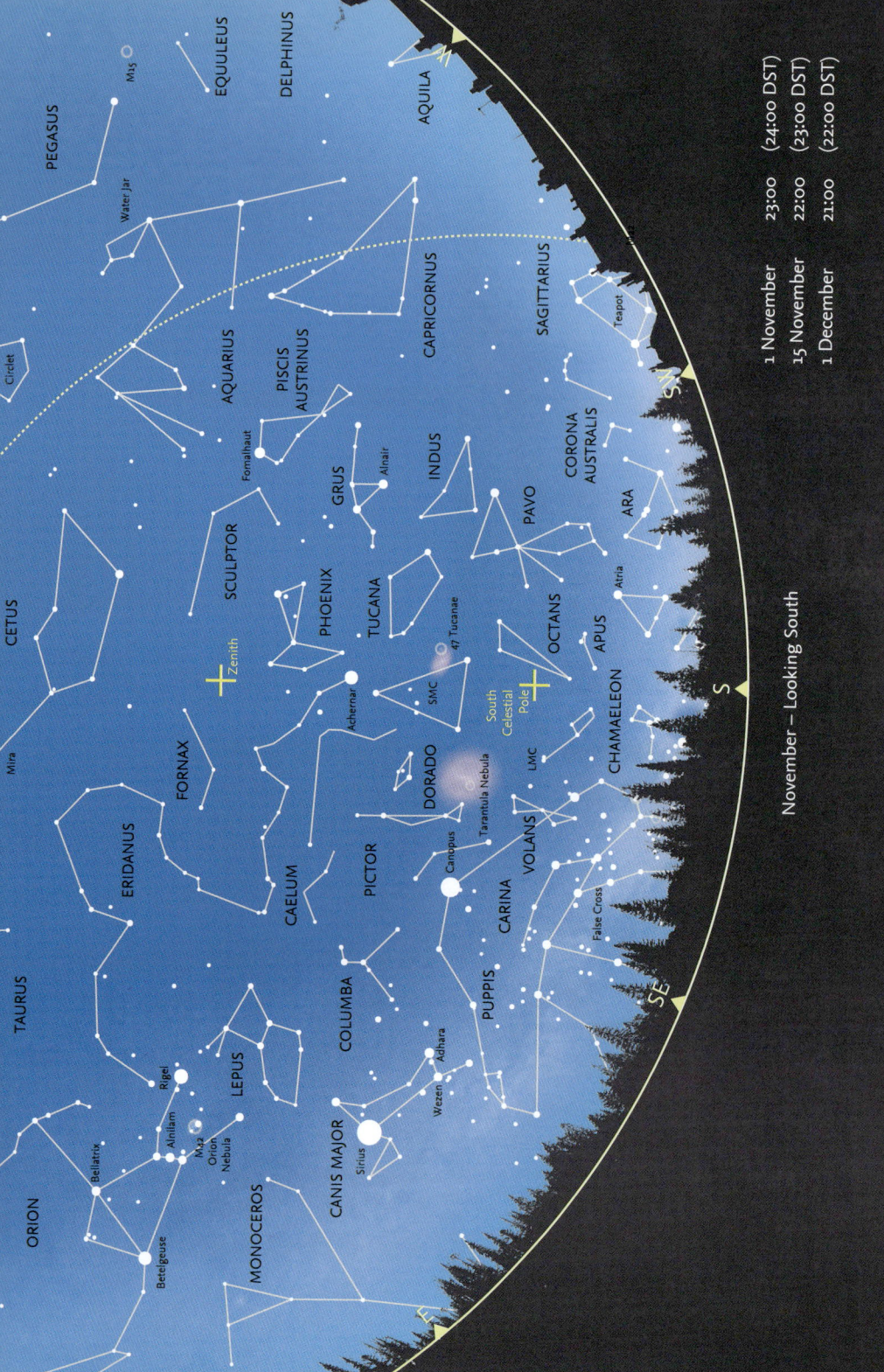

November – Looking South

Crux and the two brightest stars of **Centaurus**, **Rigil Kentaurus** (α Centauri) and **Hadar** (β Centauri), are extremely low on the southern horizon. The **False Cross** on the **Carina/Vela** border is now higher, and the constellation of **Puppis** as well as **Canopus** (α Carinae), and both the **Large Magellanic Cloud** (LMC) and the **Small Magellanic Cloud** (SMC), are clearly visible. **Achernar** (α Eridani) is halfway between the South Celestial Pole and the zenith. The whole of **Eridanus**, which starts near **Rigel** in **Orion**, is now clearly seen as it winds its way to Achernar. **Pavo** is becoming lower in the southwest and, in the west, most of **Sagittarius** is below the horizon, with **Capricornus** descending behind it. **Corona Australis** is still just visible. In the east, **Canis Major** is now clearly seen, together with the small constellations of **Columba** and **Lepus** above it.

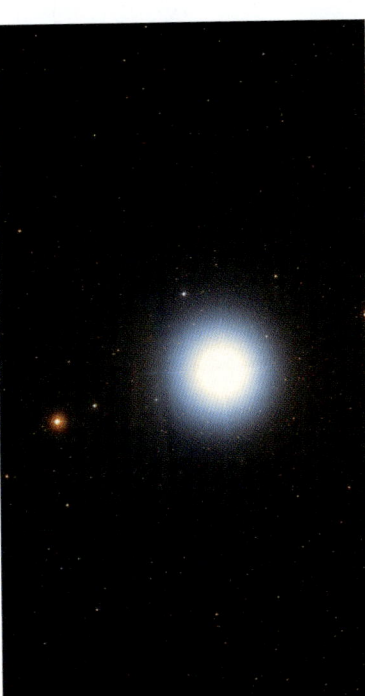

Achernar (α Eridani) at the head or end of the river Eridanus. The star is hotter than the Sun and has a pale blue colour. It is thought to be spinning rapidly and thus can be described as an oblate spheroid.

Meteors

Two meteor showers begin in September or October but continue into November. The **Orionids** (see page 89), one of the streams associated with Comet 1P/Halley, continue until about 7 November. The **Southern Taurids** (see page 83) begin on 10 September and continue until 20 November. The **Northern Taurid** shower, which began in mid-October, reaches maximum – although only with a rate of about five meteors per hour – on 12 November. The Moon is waxing crescent and setting relatively early; the meteors are best observed later in the evening. The shower gradually trails off, ending around 10 December.

Because of the location of the radiant, the **Leonid** shower is best seen from the northern hemisphere, but southern observers may see some rising from the horizon. They have a short period of activity (6–30 November), with maximum on 18 November. This shower is associated with Comet 55P/Tempel-Tuttle and has shown extraordinary activity on various occasions with many thousands of meteors per hour. The rate in 2026 is likely to be about 10–15 per hour at peak activity, however the Moon will be waxing gibbous. Conditions will be favourable after midnight. These meteors are the fastest shower meteors recorded (about 70 km per second) and often leave persistent trains.

There is a minor southern meteor shower that begins activity in late November. This is the **Phoenicids**, but little is known of the shower, partly because the parent comet is believed to be the disintegrated Comet D/1819 W1 (Blanpain). With no accurate knowledge of the location of the remnants of the comet, predicting the possible rate becomes little more than guesswork, but the rate is variable and may rapidly increase (as might be expected) if the orbit is nearby. Bright meteors tend to be quite frequent and the meteors are fairly slow. The radiant is located within **Phoenix**, not far from the border with **Eridanus** and the bright star **Achernar** (α Eridani).

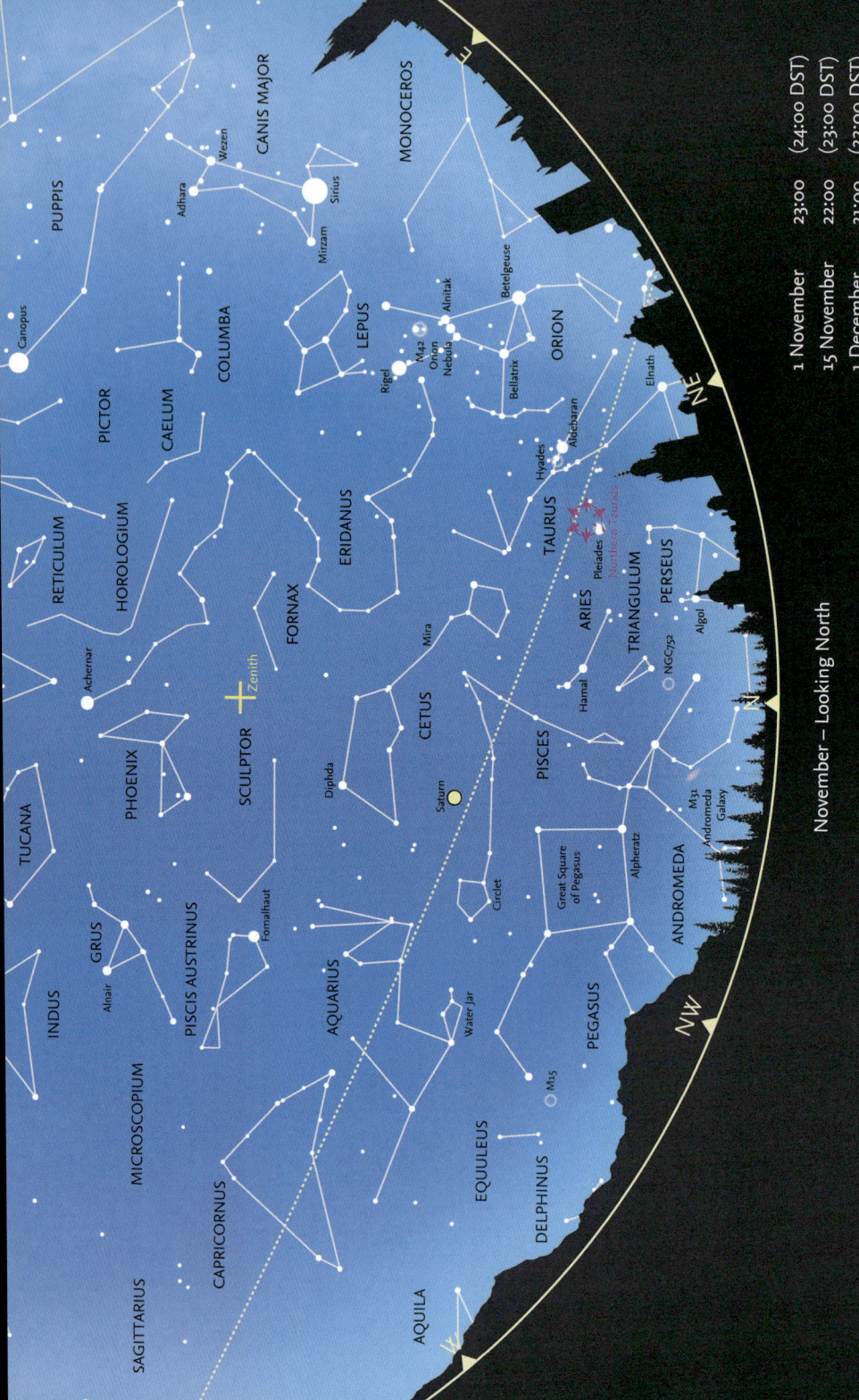

November – Looking North

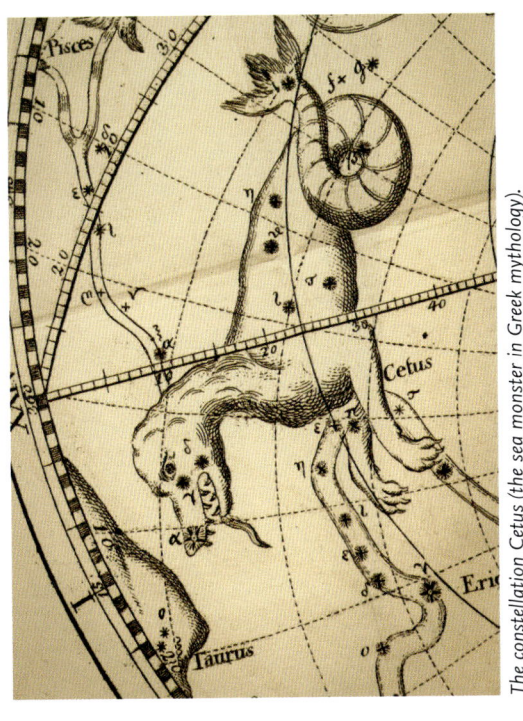

The constellation Cetus (the sea monster in Greek mythology).

Andromeda is now due north, and the **Andromeda Galaxy** (M31) has risen sufficiently to be clearly seen. The constellation of **Triangulum** lies between Andromeda and the zodiacal constellation of **Aries**. Much of **Perseus** (including the variable star, **Algol**) is now above the horizon, together with part of **Auriga**. The whole of **Pegasus**, with the **Great Square**, is clearly seen and above it the two lines of stars forming **Pisces** and the distinctive asterism of the Circlet. Higher still is the constellation of **Cetus** with the famous long-period variable star, **Mira** (o Ceti), with a typical range of magnitude 3.4 to 9.5. The whole of **Taurus** with **Aldebaran** (α Tauri), and the **Pleiades** and the **Hyades** clusters are visible in the southeast. **Orion** has fully risen in the east with **Lepus** above it. The long, winding constellation of **Eridanus** begins near **Rigel** in Orion. The faint constellation of **Monoceros** lies to the east of Orion and straddles the Milky Way. **Canis Major** and brilliant **Sirius** are even farther round towards the east. In the west, the zodiacal constellations of **Aquarius** (with its distinctive asterism of the Water Jar) and **Capricornus** are clearly seen, with **Piscis Austrinus** and bright **Fomalhaut** (α Piscis Austrini) higher in the sky. The faint constellation of **Sculptor** lies between Piscis Austrinus and the zenith.

The Moon's phases for November 2026

Last Quarter: 01, 02
First Quarter: 17, 18
Full Moon: 24, 25
New Moon: 09, 10

Day 1 – Day 30

November – Moon and Planets

The Moon

On 1 November the Moon is Last Quarter, a day later the waning gibbous Moon is 1.1°S of *Mars* (mag. 0.8) and 0.5°S of *Jupiter* (mag. -2.1). On 3 November *Regulus* is 0.8°N of the waning crescent Moon. On 7 November the Moon joins *Venus* (mag. -4.5) in *Virgo*, sitting 1.1°S of the planet. It is also 2.4°S of *Spica*. On 9 November the Moon is New; two days later the slender waxing crescent Moon moves 0.3°S of *Antares*. A First Quarter Moon dominates the sky on 17 November; a week later on 24 November the Full Moon is 0.9°N of the *Pleiades*. On 28 November the waning gibbous Moon lies 4.2°S of *Pollux*; two days later it is with Jupiter (mag. -2.2), 1.2°S of the planet and 1.1°S of *Regulus* in *Leo*. On the same day Mars (mag. 0.4) is 3.3°N of the Moon.

The Planets

Mercury is placed in *Libra* close to the Sun at sunset. It leads the Sun in the west after 4 November and moves into *Virgo* and then back into Libra as the month progresses (mag. 3.3 to 6.5, brightening to -0.7). On 20 November Mercury is at western elongation (mag. -0.5). *Venus* is visible at dawn in Virgo, receding from the Sun in the sky (mag. -4.2 to -4.9). On 10 November it lies 0.1°S of *Spica* (mag. -4.7). *Mars* is in *Leo* and creeps above the horizon around an hour before midnight (mag. 0.8 to 0.4). Mars (mag. 0.7) is 1.2°N of *Jupiter* (mag. -2.1) on 16 November. On 25 November Mars (mag. 0.6) is 1.6°N of *Regulus*. Jupiter is visible in Leo from midnight (mag. -2.0 to -2.2). *Saturn* is in *Cetus* near *Pisces*, in the sky just after sunset (mag. 0.5 to 0.6). *Uranus* is in *Taurus* from early evening (mag. 5.6), reaching opposition on 25 November at a distance of 18.4 AU from Earth. *Neptune* is in Pisces (mag. 7.7).

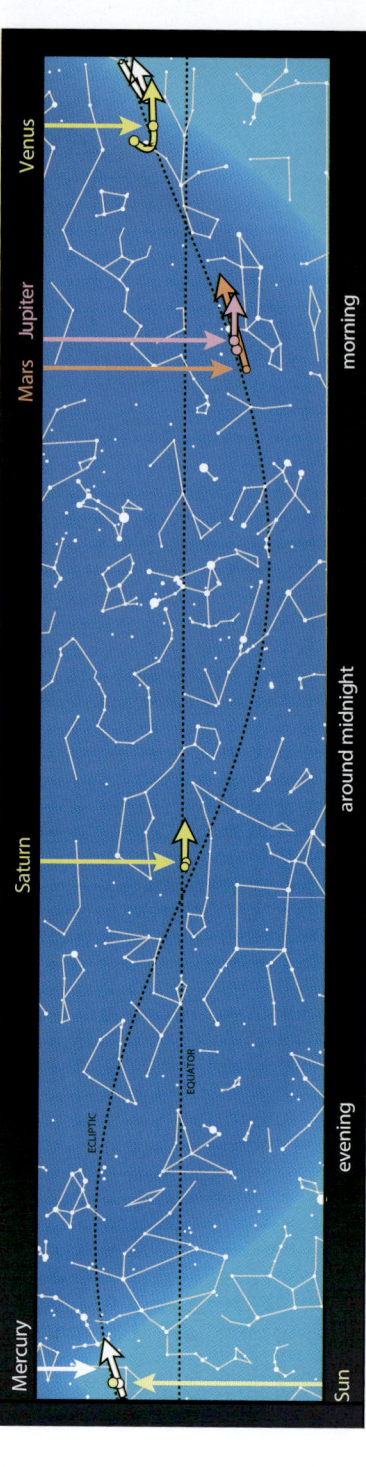

The path of the Sun and the planets along the ecliptic in November.

Calendar for November

01	20:28	Last Quarter Moon
02	14:23	Mars (mag. 0.8) 1.1°N of the Moon. Occultation visible from southern French Polynesia, Cook Islands & Pitcairn in the South Pacific Ocean
02	23:11	Jupiter (mag. -2.1) 0.5°N of the Moon. Occultation visible from Australia, Indonesia, southern India & Malaysia
03	08:40	Regulus 0.8°N of the Moon. Occultation visible from Brazil, Argentina, Peru & Bolivia
07	11:31	Venus (mag. -4.5) 1.1°N of the Moon. Occultation visible from Antarctica, Argentina, Chile & Falkland Islands
07	12:40	Spica 2.4°N of the Moon
09	07:02	New Moon
10	13:49	Venus (mag. -4.7) 0.1°S of Spica
11	03:58	Antares 0.3°N of the Moon. Occultation visible from Samoa, Tonga, American Samoa & southwestern Fiji
12		Northern Taurid meteor shower maximum
13	17:50	Moon at apogee = 405,619 km
16	06:24	Mars (mag. 0.7) 1.2°N of Jupiter
17	11:48	First Quarter Moon
17–18		Leonid meteor shower maximum
20	23:40	Mercury at western elongation (19.6°W, mag. -0.5)
24	11:18	Pleiades 0.9°S of the Moon
24	14:53	Full Moon
25	07:47	Mars (mag. 0.6) 1.6°N of Regulus
25	20:58	Moon at perigee = 359,348 km
25	22:41	Uranus (mag. 5.6) at opposition
28	01:27	Pollux 4.2°N of the Moon
30	09:18	Jupiter (mag. -2.2) 1.2°N of the Moon. Occultation visible from southern Argentina, Chile, Antarctica & Falkland Islands
30	14:35	Regulus 1.1°N of the Moon. Occultation visible from New Zealand & Norfolk Island

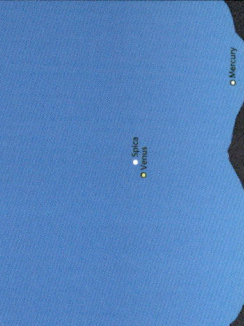

8 November • *Venus and the waning crescent Moon in Virgo close to Spica, visible shortly before sunrise.*

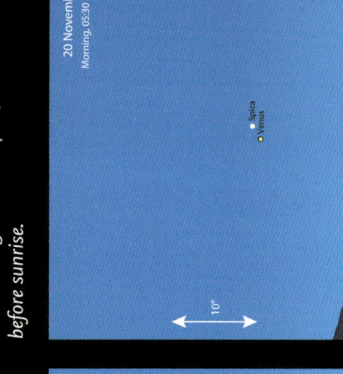

3 November • *The Last Quarter Moon with Mars and Jupiter, and Regulus close by.*

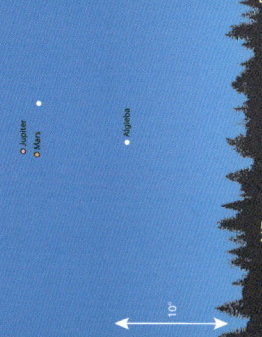

17 November • *Mars and Jupiter in conjunction in Leo, north of Regulus,*

20 November • *Mercury at western elongation, Venus close to Spica in*

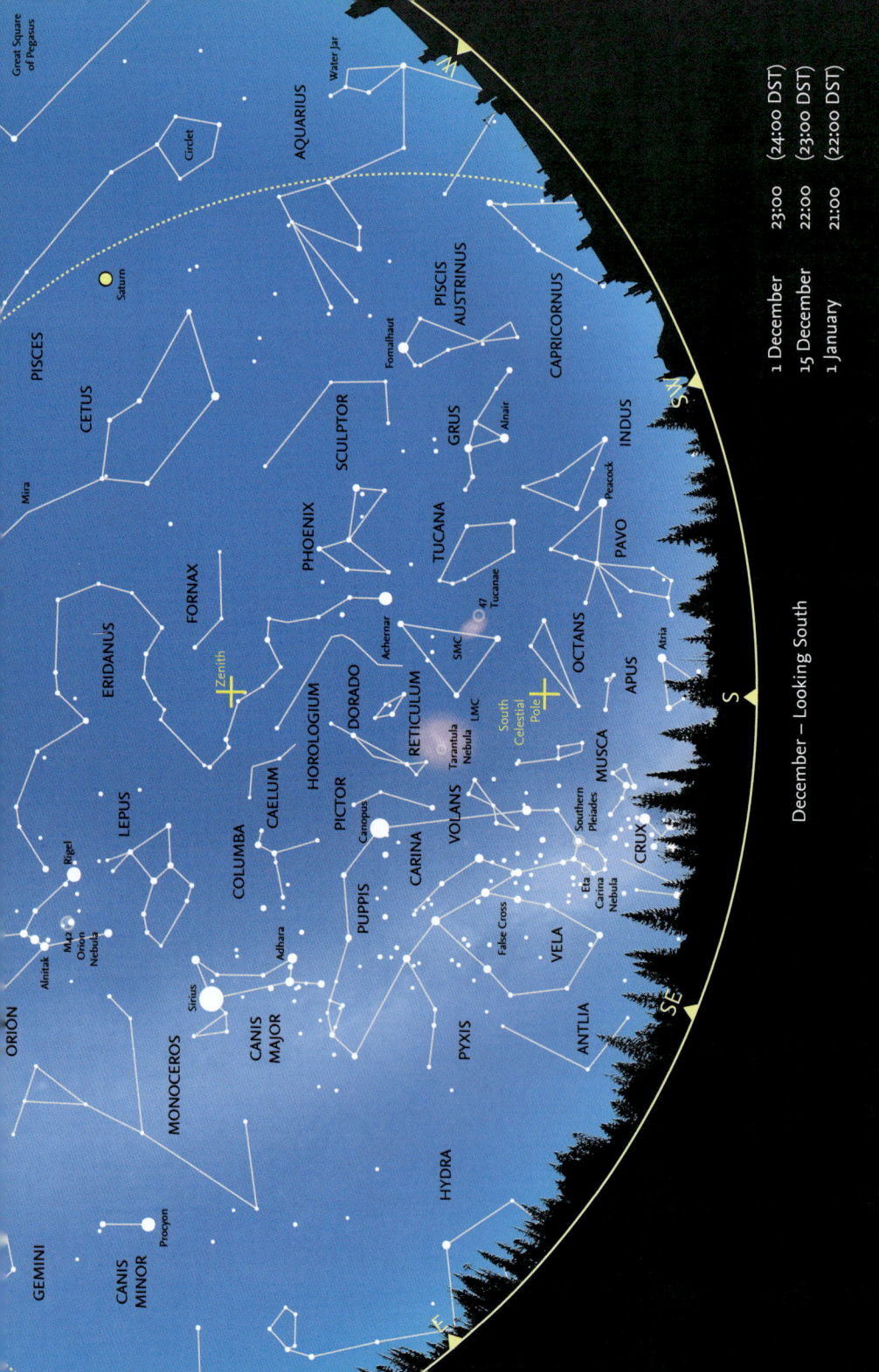

December – Looking South

Around summer solstice (21 December) southern hemisphere observers experience the longest period of daylight (midsummer's day). **Crux** and the two brightest stars in **Centaurus**, **Hadar** and **Rigil Kentaurus**, are now higher above the horizon. The **Eta Carina Nebula** and the **Southern Pleiades** are now conveniently placed for observation. Above them, the **False Cross** is clearly seen, with the whole of **Vela** and below it the constellation of **Antlia** (the air pump). **Carina** with **Canopus** (α Carinae) and **Puppis** are roughly halfway between the horizon and the zenith. **Sirius** and **Canis Major** are high in the east. In the west, **Achernar** (α Eridani) and **Phoenix** are about the same altitude as Canopus. The faint constellations of **Pictor**, **Dorado**, **Reticulum** and **Horologium** lie between them. **Pavo** with its brightest star **Peacock** (α Pavonis) is becoming low, as are the constellations of **Indus**, **Grus** and **Piscis Austrinus**. Higher still are the constellations of **Sculptor** and, near the zenith, **Fornax**. **Capricornus** is largely invisible, but most of **Aquarius** may still be seen.

Meteors

The **Phoenicid** shower continues into December, reaching its weak maximum on 2 December. The **Puppid-Velid** shower's radiant is on the border between the two constellations. The shower begins on 1 December, lasting until 15 December, with maximum on 7 December. It is a weak shower with a maximum hourly rate of about 10 meteors, but bright meteors are often seen. There is one significant meteor shower in December (the last major shower of the year). This is the **Geminid** shower, which is visible over the period 4–20 December and comes to maximum on 14 December, when the Moon is a waxing crescent; favourable conditions arise later in the evening. It is one of the most active showers of the year, with a peak rate of around 100 meteors per hour. It is the one major shower that shows good activity before midnight. The source of the Geminids is debris from an asteroid called 3200 Phaethon; most meteor showers originate from cometary debris. The Geminids are assumed to consist of denser, rocky material, they are slower than most other meteors and often appear to last longer. The brightest often break up into numerous luminous fragments that follow similar paths across the sky.

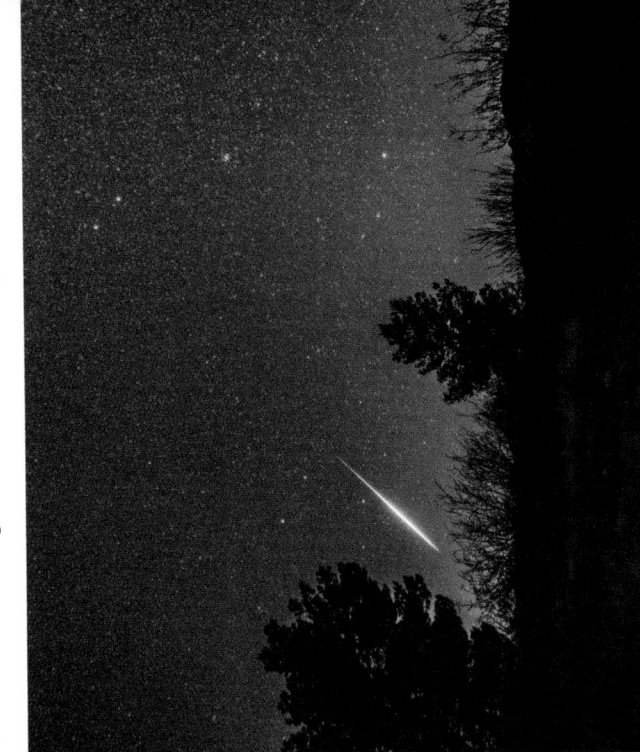

A *Geminid fireball*. The meteors can travel through the atmosphere at speeds of up to 79,000 miles per hour (127,000 kilometres per hour).

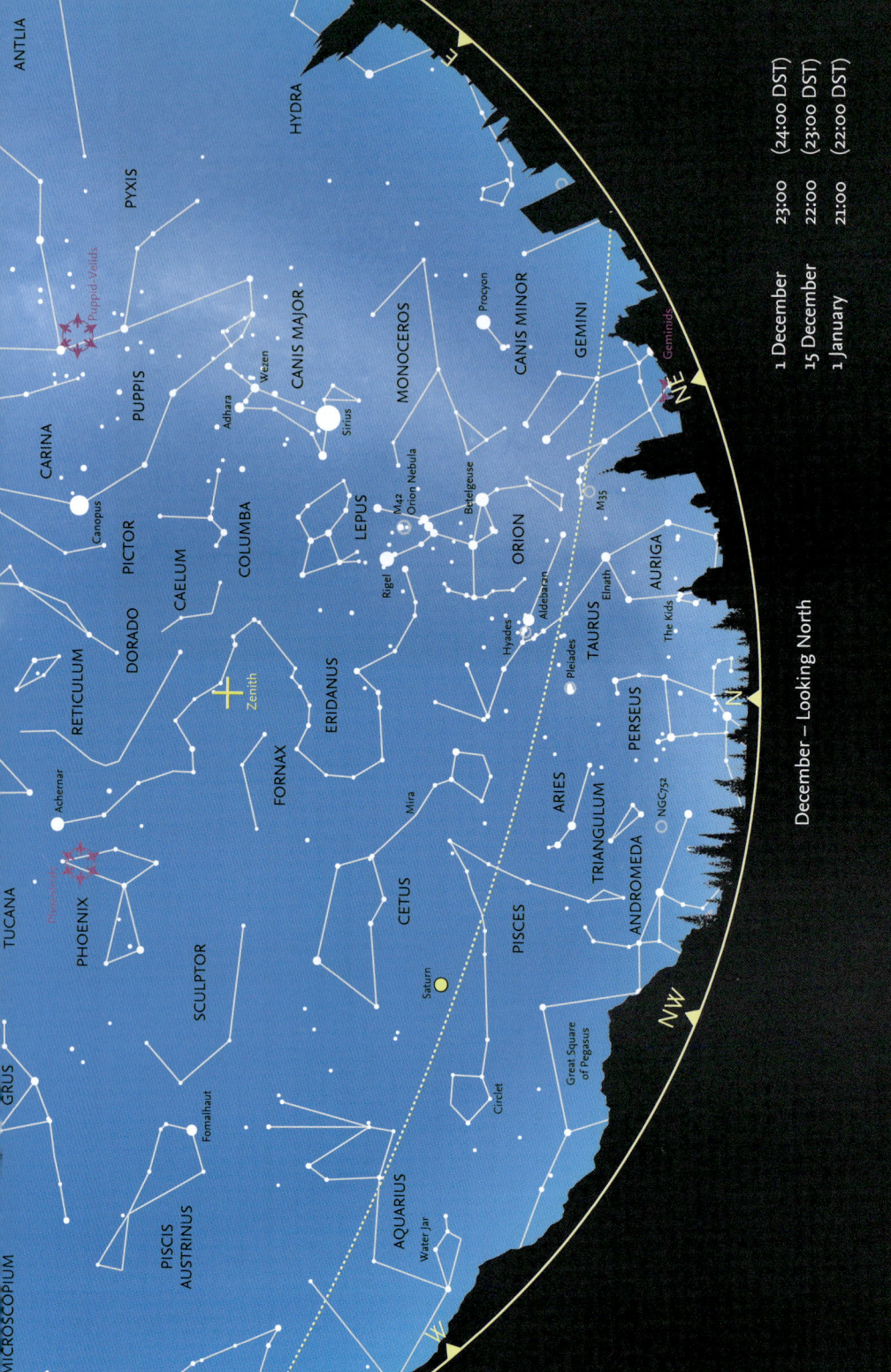

December – Looking North

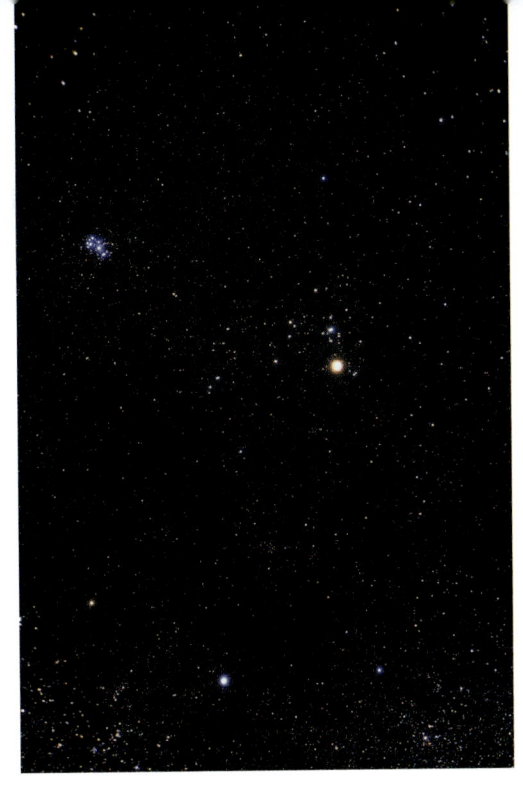

Most of *Perseus* may be seen due north with, above it, the beautiful open cluster of the *Pleiades*, the *Hyades* cluster, the orange star *Aldebaran* and the rest of *Taurus*. Farther west, *Andromeda* is very low, and the southern stars of *Pegasus* have been lost below the horizon. Above Pegasus lies the zodiacal constellation of *Pisces* and, still higher, *Cetus* and the faint constellations of *Sculptor* and *Fornax*. The famous variable of *Mira* in Cetus (see charts on pages 100 and 102) is ideally placed for observation. Towards the east, the whole of the constellations of both *Auriga* and *Gemini* are visible, although *Capella* (α Aurigae) and *Castor* and *Pollux* (α and β Gemini) are low on the horizon. *Taurus*, *Orion*, *Canis Major* (with *Sirius*, the brightest star in the sky) and *Canis Minor* (with *Procyon*) are readily visible, together with the faint constellation of *Monoceros*. *Eridanus* wanders from its start near *Rigel* (β Orionis) towards *Achernar* (α Eridani), beyond the zenith.

The constellation of Taurus contains two contrasting open clusters: the compact Pleiades, with its striking blue-white stars, and the more scattered, 'V'-shaped Hyades, which are much closer to us. Orange Aldebaran (α Tauri) forms part of the 'V'.

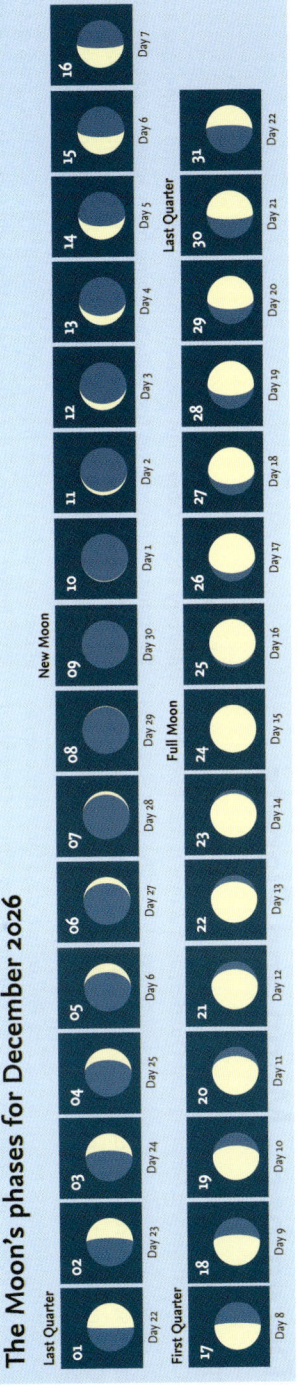

The Moon's phases for December 2026

DECEMBER 103

December – Moon and Planets

The Moon

On 1 December the Moon is Last Quarter; as it wanes it moves to 2.5°S of **Spica**. A New Moon arises on 9 December and 2 days later it is at its farthest point from the Earth (apogee), a distance of 406,421 km. First Quarter is on 17 December. The waxing gibbous Moon is 1.0°N of the Pleiades on 21 December and is Full three days later. On 25 December the waning gibbous Moon is 4.4°S of **Pollux**; two days later it moves to 1.5°S of **Jupiter** (mag. -2.4) and 1.4°S of **Regulus** in **Leo**. On 24 December the Moon makes its closest approach to Earth, with a perigee distance of 356,650 km. The Last Quarter Moon occurs on 30 December.

The Planets

Mercury sits in **Libra** then passes through **Scorpius** and **Ophiuchus** and **Sagittarius**; it approaches the Sun, switching from appearing at dawn to dusk after New Year's Day (mag. -0.7 to -1.3). **Venus** is in **Virgo** visible at dawn, it passes into Libra and reaches western elongation in the new year on 3 January 2027 (mag. -4.9 to -4.6). **Mars** stays in **Leo** for the month; it is above the horizon a few hours before midnight by New Year (mag. 0.4 to -0.1). **Jupiter** is in Leo and visible later in the night with Mars (mag. -2.2 to -2.4). On 12 December Jupiter (mag. -2.3) is 1.3°N of **Regulus**. It begins to move westwards in Leo on 13 December. **Saturn** is on the border between **Cetus** and **Pisces** and visible from sunset to just before midnight (mag. 0.6 to 0.7). On 11 December Saturn ends retrograde motion at mag. 0.6. **Uranus** is settled in **Taurus** and visible all night (mag. 5.6). **Neptune** continues in Pisces at mag. 7.7 to 7.8; it ends retrograde motion on 12 December, moving eastwards towards the end of the year.

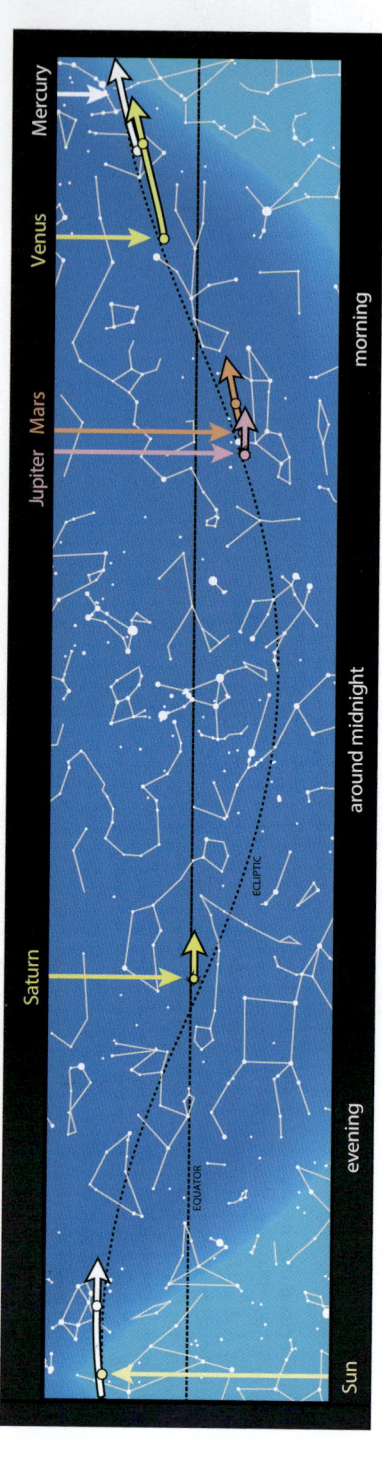

The path of the Sun and the planets along the ecliptic in December.

Calendar for December

01	06:09	Last Quarter Moon
02		Phoenicid meteor shower maximum
04	18:36	Spica 2.5°N of the Moon
07		Puppid-Velid meteor shower maximum
09	00:52	New Moon
11	06:46	Moon at apogee = 406,421 km
12	15:35	Jupiter (mag. −2.3) 1.3°N of Regulus
14		Geminid meteor shower maximum
17	05:43	First Quarter Moon
21	20:50	Winter Solstice
21	22:37	Pleiades 1.0°S of the Moon
24	01:28	Full Moon
24	08:30	Moon at perigee = 356,650 km
25	11:41	Pollux 4.4°N of the Moon
27	17:32	Jupiter (mag. −2.4) 1.5°N of the Moon
27	22:44	Regulus 1.4°N of the Moon

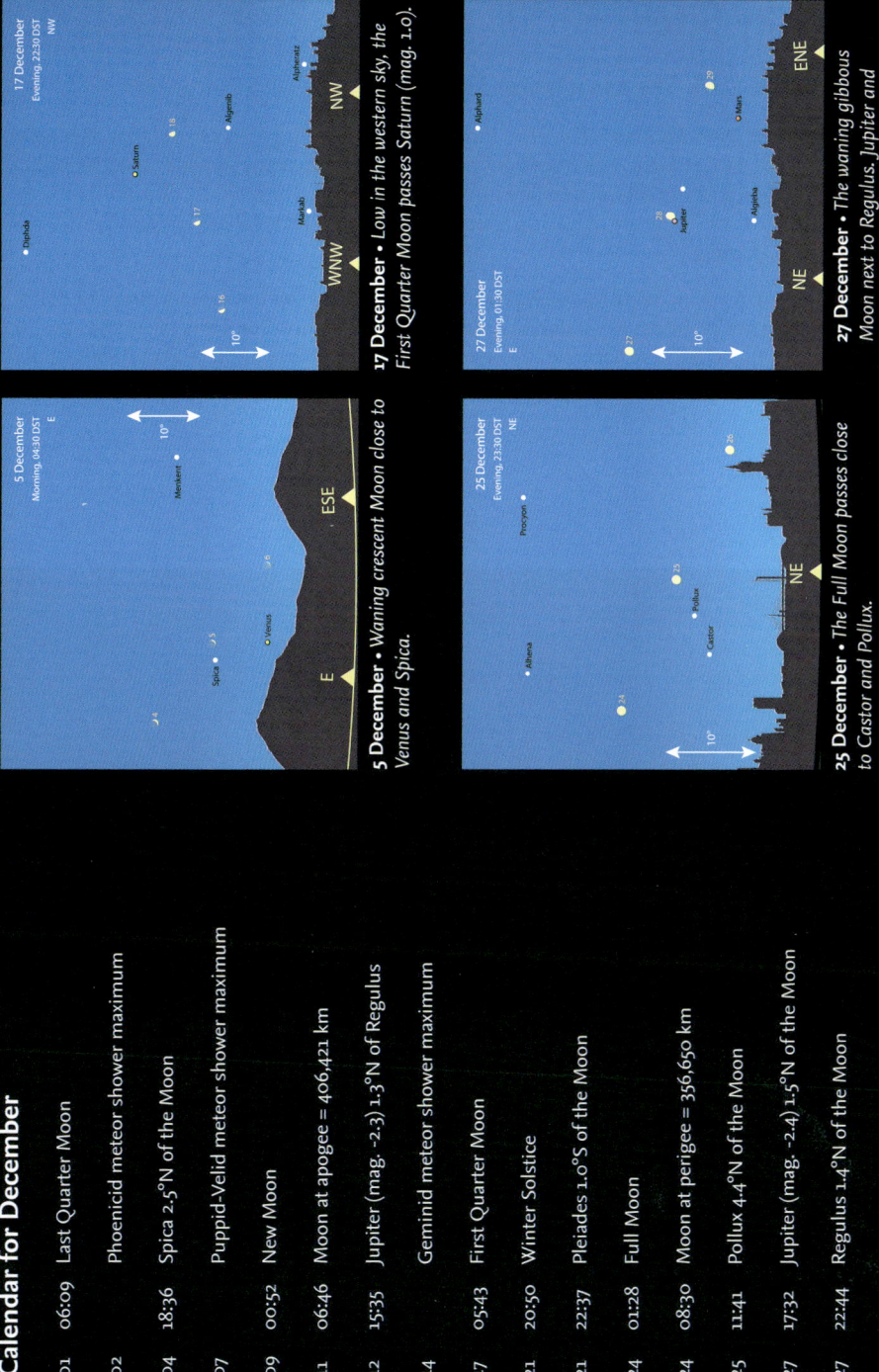

5 December • Waning crescent Moon close to Venus and Spica.

17 December • Low in the western sky, the First Quarter Moon passes Saturn (mag. 1.0).

25 December • The Full Moon passes close to Castor and Pollux.

27 December • The waning gibbous Moon next to Regulus. Jupiter and Mars are close by.

Dark Sky Sites

International Dark-Sky Association Sites

The **International Dark-Sky Association** (IDA) recognizes various categories of sites that offer areas where the sky is dark at night, free from light pollution and particularly suitable for astronomical observing. Although a number of sites are under consideration, the majority of confirmed sites are in North America. There are at least 13 sites in the southern hemisphere and the list is growing.

Details of the IDA are at: **https://www.darksky.org/**. Information on the various categories and individual sites are at: **https://www.darksky.org/our-work/conservation/idsp/**. New sites are continually being added, so check the IDA website for details. Many of these sites have major observatories or other facilities available for public observing (often at specific dates or times).

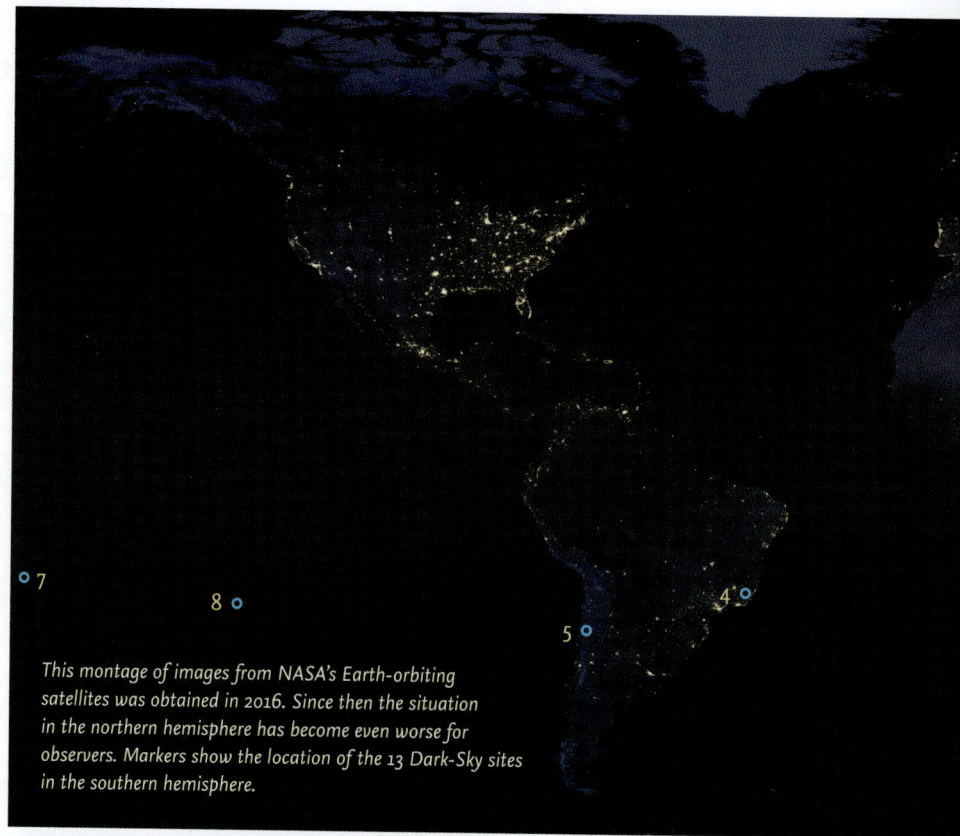

This montage of images from NASA's Earth-orbiting satellites was obtained in 2016. Since then the situation in the northern hemisphere has become even worse for observers. Markers show the location of the 13 Dark-Sky sites in the southern hemisphere.

1. *!Ae!Hai Kalahari Heritage Park* (South Africa)
2. *Aoraki Mackenzie* (New Zealand)
3. *Aotea / Great Barrier Island* (New Zealand)
4. *Desengano State Park* (Brazil)
5. *Gabriela Mistral* (Chile)
6. *NamibRand Nature Reserve* (Namibia)
7. *Niue* (New Zealand)
8. *Pitcairn Islands* (UK)
9. *River Murray* (Australia)
10. *Stewart Island / Rakiura* (New Zealand)
11. *The Jump-Up* (Australia)
12. *Wai-Iti* (New Zealand)
13. *Warrumbungle National Park* (Australia)

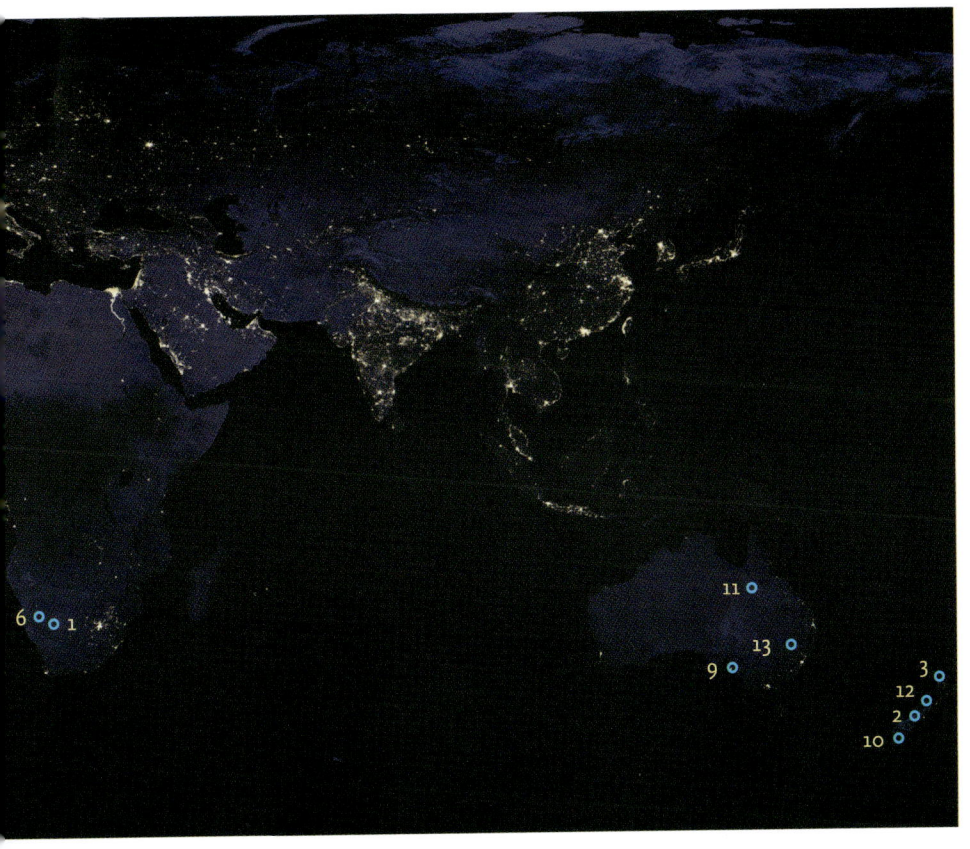

Glossary and Tables

aphelion	The point on an orbit that is farthest from the Sun.
apogee	The point on its orbit at which the Moon is farthest from the Earth.
appulse	The apparently close approach of two celestial objects; two planets, or a planet and star.
astronomical unit	(AU) The mean distance of the Earth from the Sun, 149,597,870 km.
celestial equator	The great circle on the celestial sphere that is in the same plane as the Earth's equator.
celestial sphere	The apparent sphere surrounding the Earth on which all celestial bodies (stars, planets, etc.) seem to be located.
conjunction	The point in time when two celestial objects have the same celestial longitude. In the case of the Sun and a planet, superior conjunction occurs when the planet lies on the far side of the Sun (as seen from Earth). For Mercury and Venus, inferior conjunction occurs when they pass between the Sun and the Earth.
direct motion	Motion from west to east on the sky.
ecliptic	The apparent path of the Sun across the sky throughout the year. Also: the plane of the Earth's orbit in space.
elongation	The point at which an inferior planet has the greatest angular distance from the Sun, as seen from Earth.
equinox	The two points during the year when night and day have equal duration. Also: the points on the sky at which the ecliptic intersects the celestial equator. The vernal (spring) equinox is of particular importance in astronomy.
gibbous	The stage in the sequence of phases at which the illumination of a body lies between half and full. In the case of the Moon, the term is applied to phases between First Quarter and Full, and between Full and Last Quarter.
inferior planet	Either of the planets Mercury or Venus, which have orbits inside that of the Earth.
magnitude	The brightness of a star, planet or other celestial body. It is a logarithmic scale, where larger numbers indicate fainter brightness. A difference of 5 in magnitude indicates a difference of 100 in actual brightness, thus a first-magnitude star is 100 times as bright as one of sixth magnitude.
meridian	The great circle passing through the North and South Poles of a body and the observer's position; or the corresponding great circle on the celestial sphere that passes through the North and South Celestial Poles and also through the observer's zenith.
nadir	The point on the celestial sphere directly beneath the observer's feet, opposite the zenith.
occultation	The disappearance of one celestial body behind another, such as when stars or planets are hidden behind the Moon.
opposition	The point on a superior planet's orbit at which it is directly opposite the Sun in the sky.
perigee	The point on its orbit at which the Moon is closest to the Earth.
perihelion	The point on an orbit that is closest to the Sun.
retrograde motion	Motion from east to west on the sky.
superior planet	A planet that has an orbit outside that of the Earth.
vernal equinox	The point at which the Sun, in its apparent motion along the ecliptic, crosses the celestial equator from south to north. Also known as the First Point of Aries.
zenith	The point directly above the observer's head.
zodiac	A band, streching 8° on either side of the ecliptic, within which the Moon and planets appear to move. It consists of twelve equal areas, originally named after the constellation that once lay within it.

The Constellations

There are 88 constellations covering the whole of the celestial sphere, but 4 of these in the northern hemisphere (Camelopardalis, Cassiopeia, Cepheus and Ursa Minor) can never be seen (even in part) from a latitude of 35°S, so are omitted from this table. The names themselves are expressed in Latin, and the names of stars are frequently given by Greek letters (see next page) followed by the genitive of the constellation name. The genitives and English names of the various constellations are included.

Name	Genitive	Abbr.	English name
Andromeda	Andromedae	And	Andromeda
Antlia	Antliae	Ant	Air Pump
Apus	Apodis	Aps	Bird of Paradise
Aquarius	Aquarii	Aqr	Water Bearer
Aquila	Aquilae	Aql	Eagle
Ara	Arae	Ara	Altar
Aries	Arietis	Ari	Ram
Auriga	Aurigae	Aur	Charioteer
Boötes	Boötis	Boo	Herdsman
Caelum	Caeli	Cae	Burin
Cancer	Cancri	Cnc	Crab
Canes Venatici	Canum Venaticorum	CVn	Hunting Dogs
Canis Major	Canis Majoris	CMa	Big Dog
Canis Minor	Canis Minoris	CMi	Little Dog
Capricornus	Capricorni	Cap	Sea Goat
Carina	Carinae	Car	Keel
Centaurus	Centauri	Cen	Centaur
Cetus	Ceti	Cet	Whale
Chamaeleon	Chamaeleontis	Cha	Chameleon
Circinus	Circini	Cir	Compasses
Columba	Columbae	Col	Dove
Coma Berenices	Comae Berenices	Com	Berenice's Hair
Corona Australis	Coronae Australis	CrA	Southern Crown
Corona Borealis	Coronae Borealis	CrB	Northern Crown
Corvus	Corvi	Crv	Crow
Crater	Crateris	Crt	Cup
Crux	Crucis	Cru	Southern Cross
Cygnus	Cygni	Cyg	Swan
Delphinus	Delphini	Del	Dolphin
Dorado	Doradus	Dor	Dorado
Draco	Draconis	Dra	Dragon
Equuleus	Equulei	Equ	Little Horse
Eridanus	Eridani	Eri	River Eridanus
Fornax	Fornacis	For	Furnace
Gemini	Geminorum	Gem	Twins
Grus	Gruis	Gru	Crane
Hercules	Herculis	Her	Hercules
Horologium	Horologii	Hor	Clock
Hydra	Hydrae	Hya	Water Snake
Hydrus	Hydri	Hyi	Lesser Water Snake
Indus	Indi	Ind	Indian
Lacerta	Lacertae	Lac	Lizard
Leo	Leonis	Leo	Lion
Leo Minor	Leonis Minoris	LMi	Little Lion
Lepus	Leporis	Lep	Hare
Libra	Librae	Lib	Scales
Lupus	Lupi	Lup	Wolf
Lynx	Lyncis	Lyn	Lynx
Lyra	Lyrae	Lyr	Lyre
Mensa	Mensae	Men	Table Mountain
Microscopium	Microscopii	Mic	Microscope
Monoceros	Monocerotis	Mon	Unicorn
Musca	Muscae	Mus	Fly
Norma	Normae	Nor	Set Square
Octans	Octantis	Oct	Octant
Ophiuchus	Ophiuchi	Oph	Serpent Bearer
Orion	Orionis	Ori	Orion
Pavo	Pavonis	Pav	Peacock
Pegasus	Pegasi	Peg	Pegasus
Perseus	Persei	Per	Perseus
Phoenix	Phoenicis	Phe	Phoenix
Pictor	Pictoris	Pic	Painter's Easel
Pisces	Piscium	Psc	Fishes
Piscis Austrinus	Piscis Austrini	PsA	Southern Fish
Puppis	Puppis	Pup	Stern
Pyxis	Pyxidis	Pyx	Compass
Reticulum	Reticuli	Ret	Net
Sagitta	Sagittae	Sge	Arrow
Sagittarius	Sagittarii	Sgr	Archer
Scorpius	Scorpii	Sco	Scorpion
Sculptor	Sulptoris	Scu	Sculptor
Scutum	Scuti	Sct	Shield
Serpens	Serpentis	Ser	Serpent
Sextans	Sextantis	Sex	Sextant
Taurus	Tauri	Tau	Bull
Telescopium	Telescopii	Tel	Telescope
Triangulum	Trianguli	Tri	Triangle
Triangulum Australe	Trianguli Australis	TrA	Southern Triangle
Tucana	Tucanae	Tuc	Toucan
Ursa Major	Ursae Majoris	UMa	Great Bear
Vela	Velorum	Vel	Sails
Virgo	Virginis	Vir	Virgin
Volans	Volantis	Vol	Flying Fish
Vulpecula	Vulpeculae	Vul	Fox

The Greek Alphabet

α	Alpha	ε	Epsilon	ι	Iota	ν	Nu	ρ	Rho	φ (φ)	Phi
β	Beta	ζ	Zeta	κ	Kappa	ξ	Xi	σ (ς)	Sigma	χ	Chi
γ	Gamma	η	Eta	λ	Lambda	ο	Omicron	τ	Tau	ψ	Psi
δ	Delta	θ (ϑ)	Theta	μ	Mu	π	Pi	υ	Upsilon	ω	Omega

Some common asterisms

Belt of Orion	δ, ε and ζ Orionis
Cat's Eyes	λ and υ Scorpii
Circlet	γ, θ, ι, λ and κ Piscium
False Cross	ε and ι Carinae and δ and κ Velorum
Fish Hook	α, β, δ and π Scorpii
Head of Cetus	α, γ, ξ², μ and λ Ceti
Head of Hydra	δ, ε, ζ, η, ρ and σ Hydrae
Job's Coffin	α, β, γ and δ Delphini
Keystone	ε, ζ, η and π Herculis
Kids	ζ and η Aurigae
Milk Dipper	ζ, γ, σ, φ and λ Sagittarii
Pot	= Saucepan
Saucepan	ι, θ, ζ, ε, δ and η Orionis
Sickle	α, η, γ, ζ, μ and ε Leonis
Southern Pointers	α and β Centauri
Square of Pegasus	α, β and γ Pegasi with α Andromedae
Sword of Orion	θ and ι Orionis
Teapot	γ, ε, δ, λ, φ, σ, τ and ζ Sagittarii
Water Jar	γ, η, κ and ζ Aquarii
Y of Aquarius	= Water Jar

Acknowledgements

Our thanks to Gerry Breslin and Samuel Fitzgerald at Collins, and to Ed Bloomer, Imo Bell, Catherine Muller and Sam Imperato, astronomers at Royal Observatory Greenwich.

Image Credits

pp. 16, 21, 73, 97 Shutterstock
p. 28 Dr Russell Cockman, www.russellsastronomy.com
p. 29 Christian Gloor, CC BY 2.0
p. 32 Denis Buczynski
p. 35 ESO/J. Colosimo, CC BY 4.0
p. 37 TheStarmon, CC BY 3.0
p. 41 Antonio Ferretti, CC BY-SA 4.0
p. 43 Alan Dwyer/VW Pics/Alamy Stock Photo
p. 47 Stocktrek Images Inc/Alamy Stock Photo
p. 49 Fried Lauterbach, CC BY-SA 4.0
p. 53 Tel Lekatsas, CC BY 2.0
p. 55 Roberto Mura, CC BY-SA 3.0
p. 59 Scott Roy Atwood, CC BY-SA 3.0
p. 61 cafuego from Melbourne, Australia, CC BY-SA 2.0
p. 65 Penta Springs Limited/Alamy Stock Photo
p. 67 Album/Alamy Stock Photo
p. 71 NOIRLab/NSF/AURA, CC BY 4.0
p. 77 mLu.fotos from Germany, CC BY 2.0
p. 79 Starhopper, CC BY 4.0
p. 83 Akira Fujii/ESA/Hubble
p. 85 World History Archive/Alamy Stock Photo
p. 95 Roberto Mura, CC BY 4.0
p. 101 Stocktrek Images Inc/Alamy Stock Photo
p. 103 Akira Fujii/ESA/Hubble
p. 106 National Environmental Satellite, Data, and Information Service